PERGAMON INTERNATIONAL LIBRARY
of Science, Technology, Engineering and Social Studies

*The 1000-volume original paperback library in aid of education,
industrial training and the enjoyment of leisure*

Publisher: Robert Maxwell, M.C.

AGRICULTURAL PHYSICS

THE PERGAMON TEXTBOOK
INSPECTION COPY SERVICE

An inspection copy of any book published in the Pergamon International Library
will gladly be sent to academic staff without obligation for their consideration for
course adoption or recommendation. Copies may be retained for a period of 60 days
from receipt and returned if not suitable. When a particular title is adopted or
recommended for adoption for class use and the recommendation results in a sale
of 12 or more copies, the inspection copy may be retained with our compliments.
The Publishers will be pleased to receive suggestions for revised editions and new
titles to be published in this important International Library.

Other Titles of Interest

PLATE I. Clay crumbs after exposure for 30 min to rainfall of rate about 2·2 in. hr⁻¹, drop diameter 5·1 mm, and impact velocity 4·7 m sec⁻¹ (height of fall 1·25 m). Top to bottom are samples of Willalooka subsoil clay fraction (85 per cent illite and 15 per cent kaolinite), pure kaolinite, and Wyoming bentonite (montmorillonite). In preparation clays were calcium saturated and washed until chloride molarity in the percolate was less than 10⁻⁴. Clays were free of organic matter.

AGRICULTURAL PHYSICS

C. W. ROSE, B.Sc., B.E., A.Inst.P., Ph.D.

Division of Land Research, C.S.I.R.O., Australia;
lately of Makerere University College, Uganda

PERGAMON PRESS

OXFORD · NEW YORK · TORONTO
SYDNEY · PARIS . FRANKFURT

U.K.	Pergamon Press Ltd., Headington Hill Hall, Oxford OX3 0BW, England
U.S.A.	Pergamon Press Inc., Maxwell House, Fairview Park, Elmsford, New York 10523, U.S.A.
CANADA	Pergamon of Canada, Suite 104, 150 Consumers Road, Willowdale, Ontario M2J 1P9, Canada
AUSTRALIA	Pergamon Press (Aust.) Pty. Ltd., P.O. Box 544, Potts Point, N.S.W. 2011, Australia
FRANCE	Pergamon Press SARL, 24 rue des Ecoles, 75240 Paris, Cedex 05, France
FEDERAL REPUBLIC OF GERMANY	Pergamon Press GmbH, 6242 Kronberg-Taunus, Pferdstrasse 1, Federal Republic of Germany

First edition 1966

Reprinted 1969, 1979

Library of Congress Catalog Card No. 66-18398

Printed in Great Britain by A. Wheaton & Co. Ltd, Exeter
ISBN 0 08 011884 4

Contents

Preface

IN THIS book I have attempted to discuss a range of topics in agriculture and environmental biology where a physical understanding of processes is necessary. The title *Agricultural Physics* is not intended to imply a special kind of physics, but a consideration of agricultural problems, and some aspects of the environment and water relations of plants, from a physical point of view. Whilst the book reflects the very rapid increase in the amount of research in agricultural problems and environmental biology involving a physical insight, the need for the point of view of other disciplines in this field of enquiry has not been ignored.

A wide range of subject matter is covered, and in a text of this size considerable selection and omission was necessary. Three general principles have been borne in mind in making this selection. Firstly, particular attention has been given to clarifying fundamental concepts and processes. Thus, for example, the concept of the total potential of water and its components, which is of basic importance in understanding water movement in soil, plant or atmosphere, receives a full discussion. Secondly, subject matter is limited to topics in which physics has made a significant contribution. Thus the experimental aspects of crop water use studies, for example, receive fairly detailed attention. Finally, where there is a common interest, an attempt has been made to cross barriers between literatures in different disciplines, a sometimes frustrating but often rewarding enterprise.

It is appreciated that there may be readers whose interests lie chiefly in one or more of the fields of enquiry covered by this book. It may be useful therefore to indicate that sections of the text are sufficiently self-contained that at least the following four groups of chapters may be read independently without undue difficulty being caused by lack of context:

(i) Chapters 1 to 3 are concerned solely with the physical

     environment of agriculture and provide a background to the literature on the micrometeorology of crops and single plants.

(ii) For a person chiefly interested in soils Chapters 4 to 6 are quite self-contained.

(iii) A great number of agriculturalists are concerned with some aspect of crop water use. Chapters 5 to 7, and possibly Chapter 8 would be of most relevance in this connection.

(iv) For physiologists concerned with plant water relations Chapters 5 and 8 would be those of closest interest.

It is appreciated that the physical and mathematical equipment of many research workers and students of agriculture and biology is not beyond the secondary level of education. Where the subject matter requires physical understanding beyond this level an endeavour has been made to provide this background. This is necessary, for example, in understanding the physical environment of agriculture (discussed in Chapters 1 to 3) where the gap between the secondary standard and that of current research literature and more advanced and specialized texts is particularly noticeable.

In providing this background the presentation has been kept as simple as possible. Whilst not claiming that this always makes easy reading, it is hoped that it will reduce the frustration of discovering that understanding the subject matter must await the study of other texts.

It is a pleasure to acknowledge my debt to Prof. H. Birch and P. W. Webster, my former colleagues at Makerere University College, Uganda, who kindly read the entire book in draft form, making many valuable suggestions. I am also most grateful to Dr. J. R. Philip and other members of the C.S.I.R.O. for discussion on the later chapters. For her careful assistance I owe much to my wife. For permission to reproduce diagrams from publications I wish to thank the Royal Meteorological Society (Figs. 2, 12, 21); Dr. R. O. Slatyer and Mr. I. C. McIlroy (Fig. 10); the Editor, *C.S.I.R.O. Journals* (Figs. 11, 13, 45); the Institute of Physics and Physical Society (Fig. 22); the Editor, *Journal of Soil Science* (Figs. 26, 38 and Plate I); the Editor, *Soil Science* (Fig. 40); and the Editor, *Agronomy Journal* (Fig. 50).

Canberra, Australia   C. W. ROSE
1965

Important Symbols

A	area, psychrometric constant, or exchange coefficient
A_o	one standard atmosphere pressure
A_o'	gas pressure
B	overburden pressure
C	differential water capacity, or amount of a conservative quantity stored in a particular volume
C_7	cation exchange capacity at pH7
D_l	isothermal soil water diffusivity
D_o	molecular diffusion coefficient of water vapour through air
ΔD	increase in surface water detention
E	evaporation flux density
$\mathcal{E}$	radiant emittance (or emissive power)
F	flux density of radiation
F_o	radiation flux density at the earth's surface normal to the sun's radiation with cloudless skies, or upward flux density of some conservative quantity at the soil surface
F_s	force on unit mass of water in the direction s
$\bar{F}$	mean flux density of any conservative quantity
G	heat flux density into the ground, or pneumatic potential
H	sensible (or non-latent) heat flux density into the atmosphere from the ground, or hydraulic head
H_0, H_1	horizontal flux densities of some conservative quantity
I	intensity of radiation, or equivalent depth of applied irrigation water
K	hydraulic conductivity of soil
K_m	transport constant for momentum
K_h	transport constant for heat
K_t	transport constant or transfer coefficient
K_w	transport constant for water vapour
L	latent heat of vaporization of water
M	matric or capillary potential

M_a	molecular weight of dry air
M_w	molecular weight of water vapour
ΔM	increase in soil water storage in a soil volume
N	possible hours of sunshine per day
O	osmotic potential
P	precipitation, or pressure potential
Q	quantity of heat
R	pore radius
R_i	Richardson number
R_L	net flux density of long-wave radiation emitted by the ground surface
R_N	total net radiation flux density, including both long and short wavelengths
R_S	flux density of short-wave radiation received from the sun and sky on a horizontal surface at ground level
R_u	universal gas constant
S	submergence (or piezometric) potential, or net surface run-off
T	temperature in degrees absolute or Kelvin (°K), or time interval
T_a	mean air temperature (°K)
T_{dp}	dew point temperature
T_o	surface temperature
T_w	wet-bulb temperature
U	drainage beyond depth to which ΔM is calculated in water conservation equation
V	volume
W	work expended
Z	gravitational potential
a	pore radius
b	as a suffix refers to black body radiation
c	fraction of the sky covered by cloud, or a conservative property per unit mass of fluid, or specific heat of soil
c_p	specific heat of air at constant pressure
d	diameter, zero-plane displacement
e	water vapour pressure

e_o	surface water vapour pressure
e_s	(and e_a in Chapter 3) saturation vapour pressure of water at air temperature
e_w	saturation vapour pressure at the wet-bulb temperature
g	acceleration due to gravity, or gram
h	height of water column (or soil water suction expressed in this unit), or heat transfer coefficient
h_r	relative humidity (as a fraction)
k	thermal conductivity, or von Kármán constant
n	actual hours of sunshine per day, or number of stomata per unit leaf area
p	gauge pressure of soil water (i.e. pressure of soil water measured from the gas pressure acting on the soil water)
p_a	atmospheric pressure
p_o	external gas pressure measured from standard atmospheric pressure
p_w	absolute pressure in soil water
q	specific humidity
q_v	water vapour flux density
r	pore radius, component resistance to the transpiration stream
s	unloaded suction, or distance in the s-direction
t	time
u	component of wind velocity in the x-direction
u'	eddy velocity corresponding to velocity u
$\bar{u}$	mean velocity corresponding to velocity u
u_*	friction velocity
v	speed of propagation of temperature wave, or volume of water crossing unit area per second
v_t	terminal velocity
w	water content on a mass basis, or component of wind velocity in the upward (z-) direction
w'	eddy velocity corresponding to velocity w
x	mixing ratio of moist air, or distance in the x-direction
y	distance in the y-direction
z	distance in the z-direction

z_o	roughness length
a	absorptivity (absorption coefficient), or a load partition coefficient
β	Bowen ratio
γ	a ratio appearing in the wet- and dry-bulb psychrometer equation
Γ	adiabatic lapse rate
δ	small finite difference, or boundary layer thickness
Δ	small finite difference, or slope of the vapour pressure curve
ε	porosity
ζ	zeta potential
θ	volume fraction of any component, taken to be water if used without suffix; also angle or potential temperature
θ_s	saturation ratio
κ	thermal diffusivity (or thermometric conductivity)
λ	wavelength
μ	micron, also dynamic viscosity
ν	kinematic viscosity
π	osmotic suction, or the ratio of circumference to diameter of a circle
ρ	reflectivity (or reflection coefficient), or the density of water
ρ_a	moist air density
ρ_b	bulk (or apparent) density of soil
ρ_a'	dry air density
ρ_m	density of mercury
ρ_v	density of water vapour
σ	Stefan–Boltzmann constant, or surface tension of water
τ	matric (or soil water) suction, transmissivity (or transmission coefficient), or shearing stress
φ	flux
Φ	hydraulic potential
Ψ	total potential of water
Ψ_v	total potential of water vapour
ω	solid angle, or angular frequency

CHAPTER 1

The Physical Environment of Agriculture: Part I

1.1 Why Physics in Agriculture?

The baffling problem of how plants can grow without any immediate obvious source of food supply is still being unravelled. Investigations stretching over the last few hundred years have led to an understanding of many processes involved in plant growth, but this increased understanding has at the same time led us on to ask still further questions, in the way that scientific investigation always seems to do. It is not surprising that soil was first thought to be the sole supplier of food for plants. How it came to be realized that plants "feed" chiefly by absorbing carbon dioxide as a gas from the atmosphere in the presence of light, synthesizing more complex products of higher chemical potential energy, is one of the fascinating stories of scientific discovery. Chemists as well as scientists interested in plants and soils were involved in this discovery of *photosynthesis*, thus illustrating the importance of co-operation between the sciences in the understanding of biological problems. Experience that physics can also be a significant contributor in this quest provides the general basis of this book.

Of course, those early thinkers who imagined the soil might be the only source of plant food were not completely on the wrong track. For plants to be able to photosynthesize effectively they require not only support, but also a supply of water and a range of nutrients or chemicals in a suitable form. Whilst these requirements can be provided in other ways, they are normally supplied by or stored in the soil, which is also host to a vast population of living organisms intimately connected with the plant's growth.

Thus we see that the supply of food for plants depends on the condition and constituents of the plant's environment, comprised of both the lower layer of the atmosphere and the upper layer of the earth's crust. In broad terms the environment of plants is determined by the climate and the weather, which is dependent on processes involving large regions of the atmosphere. However, the earth's surface and its vegetation very significantly modify this large scale climate, giving rise to the local climate, or *microclimate* in which plants live. This term microclimate or *microenvironment* may be taken to refer not only to the climatic factors in the air layer near the ground, but also to those at and beneath the soil surface. Above ground the term implies a useful but by no means rigid distinction from the scale of interest in meteorology; whilst below ground a similar difference from the depths of common geophysical interest is also suggested. The ground surface plays a unique role in partitioning radiant energy into various components and in dividing rainfall into infiltration, evaporation and run-off. Partly because of this there are remarkably rapid changes in climatic variables near to it. For example, temperature gradients of the order of degrees per metre just above ground, and a degree per centimetre just below it are common; whereas temperature gradients above the immediate influence of the ground surface may be of the order of 1°C per 100 m.

As is described in texts on plant physiology, and as shall be briefly mentioned from time to time in this book, physical aspects of the environment can affect the entire growth and development of a plant, from germination to maturation. We have already mentioned the vital role of sunlight and carbon dioxide concentration in the air, and of water and nutrients in soil. Hence the importance of understanding the microenvironment of plants or crops, the processes involved in its change, and the continual interaction between plants and their environment. It is for this reason that we will consider such environmental factors as radiation, temperature, humidity and soil water content. We will discuss how heat and carbon dioxide exchanges and evaporation take place and the factors affecting them, how water moves in soils and through the plant, and something of why soils behave as they do. From the sections which follow it will become apparent that to gain an understanding of such processes physical insights, among others, are required. However, the following general discussion of the interaction between vegetation and its environment gives an illustration of the significance of physics in understanding processes of importance in agriculture.

VEGETATION–ENVIRONMENT INTERACTIONS

Following Penman[1] *assimilation* or the photosynthesis of carbohydrates may be regarded as the fundamental plant environment interaction. The multitude of other processes necessary for growth require energy which is obtained by *respiration*, the oxidation of more complex carbohydrates to simpler components with the release of energy. Over a period of days respiration requires the breakdown of only a fraction of the carbohydrate synthesized, but this fraction can vary considerably during the life of a plant. The difference between daily periodic assimilation and this continuous respiration is referred to as *net assimilation*.

Let us now consider some of the factors that limit the net assimilation rate in crops. A fundamental biological restriction is given by the nature of the relation between the net assimilation rate of a particular plant leaf and incident light intensity, which is of the type shown in Fig. 1. At higher light intensities the net assimilation rate remains almost constant, indicating that the rate is limited not by light intensity but perhaps by internal biochemical processes or the supply of carbon dioxide from the atmosphere. The rate of supply of carbon dioxide for photosynthesis depends both on its concentration just outside the leaf and the various resistances to diffusion which the carbon dioxide molecules experience in their passage from the outside air to the location of their photochemical reaction. These diffusive resistances depend on the physical behaviour of gas molecules. The concentration of carbon dioxide just outside a particular leaf in a crop canopy depends primarily on the efficiency of mixing of the air of that particular layer with that above the crop, this mixing efficiency depending on the turbulent character of the air flow above and through the crop. Another source of carbon dioxide, which is commonly of secondary magnitude, is that produced in the soil by respiration of plant roots and microbial decomposition

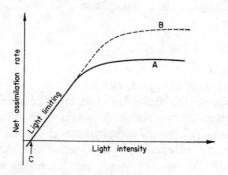

FIG. 1. Illustrating the type of dependence of net assimilation rate on light intensity with carbon dioxide supply at one level (*A*), and at a higher level (*B*).

processes. If carbon dioxide supply is limiting net assimilation rate at higher light intensities for the leaf whose response is shown by curve *A* in Fig. 1, then curve *B* illustrates the type of response to an increase in this supply.

At lower light intensities net assimilation rate is almost proportional to light intensity (Fig. 1), and falls to zero at some finite light intensity shown by point *C* on the abscissa. At this intensity synthesis of organic matter and its breakdown by respiration take place at equal rates. Point *C* is thus called the *compensation point*.

For whole plants, whilst some leaves may be *saturated* with respect to light, as illustrated by the high light intensity end of curves *A* and *B*, light can still be limiting photosynthesis for leaves partially or wholly shaded by others above them. In particular the leaf area per unit area of soil below it (called the *leaf area index*) often increases up to fruiting or senescence causing progressive reduction in light at lower levels in the canopy. This leads to a complex and important interaction between the structure of plants or plant communities and the physical properties of the incident light, illustrated by the analysis of Davidson and Philip.[2] An example of this interaction is in competition for light between adjacent species. Physical properties of light which are significant for photosynthesis in whole plants are its intensity, quality or wavelength distribution, the proportion of direct to diffuse or scattered sunlight, and their directional properties. These properties and their significance will be discussed in section 1.2, but we may note that the importance of light quality, or its spectral distribution, lies in the fact that not all wavelengths are photochemically effective.

By making reasonable assumptions concerning the factors mentioned so far, Loomis and Williams[3] have calculated a maximum rate of production of dry matter for an assumed radiation level. This calculation assumes no growth limitation due for example to lack of water, of nutrients, and to physiological ageing or attack by pests. De Wit[4] carries through a similar but more detailed analysis of potential photosynthesis, using the type of relationship shown in Fig. 1 to describe plant response. This type of approach clarifies the factors affecting the upper limits to assimilation.

Let us now turn to consider some of the interactions between photosynthesis and transpiration. Any difference between transpiration and water uptake through the roots of a plant over a period of time must result in a change of water stored in the plant. Decreases in such storage cause a reduction in the pressure of water in the plant, an analogy being the removal of water from a saturated sponge by the application of a suction, or pressure less than atmospheric. It is changes in this water pressure in a plant that partly control the opening and closing of *stomata*, the apertures through which diffusion takes place for both the carbon dioxide used in photosynthesis, and water vapour in transpiration. Hence there is a very direct but complicated interrelation between transpiration, photosynthesis and the water supply in the soil. The interrelation is complicated partly because it is essentially dynamic in character, depending on the rates of the flows associated with each process. For example, when relatively high evaporation rates are possible, as around midday, plants can be found wilted even with water apparently freely available to the roots. It is simply that rate of loss of moisture has exceeded its intake rate through the roots, and wilting is due to the loss of water pressure or turgor accompanying the net loss of water. Since such water stress often occurs during periods of maximum radiation, and thus of maximum potential photosynthesis, such stress is a common cause of yield reduction. However, if transpiration rate does not exceed the rate of root extraction there will be no water stress, and so both transpiration and photosynthetic rates will be limited by factors other than water supply, some of which have already been mentioned for photosynthesis. But what factors will then limit transpiration? Considerable research on this question has shown that the transpiration of an actively growing low level crop of large extent, completely covering the ground and not short of water is very largely controlled by physical meteorological factors, and is referred to as *potential evapotranspiration*. This definition was adopted at a meeting on physics in agriculture in 1955 (Penman[5]). Under these conditions for potential evapotranspiration a maximum fraction of the radiant energy absorbed is used in providing the latent heat which is one necessity for evaporation to take place. Another requirement for evaporation to continue is that the evaporated moisture must be removed, otherwise the air would become saturated with water vapour. If a crop is not "low and level", as stipulated in the definition of potential evapotranspiration, but consists of projecting plants around which air can flow, the removal of evaporated water is facilitated and evapotranspiration can then exceed this "potential" rate, as can also occur if the other restrictions mentioned in defining potential evapotranspiration do not hold.

The effect of vegetation in modifying air movement within and above it is a particular example of the general fact that the immediate environment of a plant community is the result of interaction between its own properties and the past and present weather conditions (Raschke[6]). The reduction of air speed in the vegetation zone obviously depends on its density, its structure, and, with row crops, on wind direction. With wheat, a crop that deflects noticeably with wind, Penman and Long[7] found a different type of wind profile in calm and in moderate wind. Such profiles, normalized to make the average velocity at the highest recording level unity, are compared in Fig. 2. Evidently, air can move more easily between the palisades in calmer conditions. In contrast the effect of the moderate wind was partially to "seal" the crop, as indicated by the steep velocity gradient on the upper part of the crop.

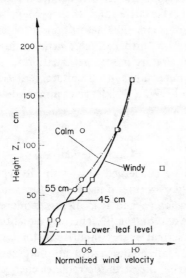

Fig. 2. Normalized mean wind profiles in calm and moderate wind. Crop height in calm, 55 cm; in wind, assumed 45 cm. (After Penman and Long.[7])

The big reduction of air movement within crops is of great importance to the plant, to insect life, and often to the spread of plant disease, directly or indirectly. Since mixing of the air is restricted, gradients of humidity and temperature greater than in the free air above the crop will also tend to exist. Geiger[8] gives examples of this. However, in general the microclimate within vegetation is much more moderate than over bare ground. Vegetation allows the solar input to be absorbed over a depth and not just at a surface. With bare soil evaporation is usually restricted far more quickly than with plants, and restricted evaporation means an increased heating of the air.

Some physical factors of the environment such as day length and temperature may have a less direct effect on growth than radiation, and they may affect *development* more than *growth*. In this context growth may be taken simply as an increase in plant material, and development as a sequence of phases, such as germination, the vegetative and the reproductive phases. In particular some environmental factors appear to have the effect of a "trigger action" for further development, such as the effect of moisture and temperature in seed germination.

Because of its importance in agriculture, radiation will now be discussed more fully. Van Wijk[9] gives a more detailed presentation of this topic.

1.2. Radiation

Here we shall be concerned only with the type of radiation emitted by a body by virtue of its temperature. Such radiation is called *thermal radiation*, and it is discussed more fully by Van Wijk,[9] Geiger[8] and Sutton.[10] Thermal radiation from the sun is the ultimate thermal energy input term for the earth and its atmosphere. Since all bodies above absolute zero temperature radiate continuously, the earth too radiates energy. Although the atmosphere acts somewhat like a blanket, much reducing the escape of the earth's radiation into space, that fraction which is lost must be almost exactly balanced on average by absorption of radiation from the sun. If this were not so the earth as a whole would be cooling (or heating) at a much greater rate than it is. Before discussing this continuous radiant exchange further, some definitions and laws we will need concerning radiation are given. It should be mentioned that our interest in thermal radiation centres in its emission and absorption and lies little in its optical properties such as rectilinear propagation, though these are assumed.

DEFINITIONS

The amount of radiant energy per unit time emitted, received, or transmitted across a particular area, is called the *radiant flux* φ. This flux divided by the area across which the radiation is transmitted is called the *radiant flux density F*. Thus $F = (d\varphi/dA)$, where A represents area. However the word flux is not infrequently used as synonomous with flux density. The radiant flux density emitted by a source is often called the *radiant emittance*, or *emissive power $\mathscr{E}$*, of the source. Measurements show that the radiant emittance of a particular body depends on the nature of its surface as well as increasing with the temperature of the body. Units of F and $\mathscr{E}$ can be gram calories per square centimetre per minute, relations with other units being given in the Appendix.

Consider an observer at point O viewing an isolated cloud in the sky. Of the total radiant flux crossing a horizontal surface at O, some will come from the cloud. Each straight line that could be drawn from O to intersect the cloud lies within the *solid angle* ω subtended at O by the cloud. Solid angle is dimensionless, the unit of solid angle being one *steradian*. Imagine a sphere of unit radius drawn with point O as centre (Fig. 3). Then the solid angle ω subtended by cloud C at O is numerically equal to the area intercepted on this unit sphere by all straight lines from O which pass through the cloud. The *radiant intensity I* in a particular direction is defined as the radiant flux per unit solid angle, i.e. $I = (d\varphi/d\omega)$. Whilst the relation between flux density F and intensity I depends on the magnitude of I in all directions of the hemisphere above the area for which F is defined, if I is the same in all directions it may be shown that: $F = \pi I$. To avoid confusion it should be noted that especially in the non-physical literature the term intensity is frequently used loosely, and often synonomously, with radiant flux density.

The *absorption coefficient a*, also referred to as the *absorptivity*, is the fraction of incident radiation intensity absorbed by a body. The *reflection coefficient ρ* and *transmission coefficient τ* are similarly defined as the fraction of incident radiation intensity

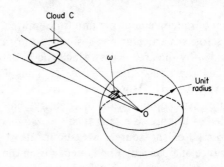

FIG. 3. A figure used in defining solid angle.

reflected and transmitted respectively. From the principle of energy conservation:

$$a + \rho + \tau = 1.$$

The magnitude of contributions to the flux φ of radiation depends strongly on the wavelength λ of the radiation. All the other properties defined above are similarly dependent on the radiation wavelength, but we shall mostly be concerned with total values integrated over all wavelengths.

Some fundamental ideas and relations

A body which absorbed all radiation falling on it might be called a *black body*. Such a body can be well approximated by a cavity enclosed by an opaque wall maintained at a uniform temperature, and pierced by an opening of dimensions small compared to those of the cavity. This is a black body since only a very small fraction of radiation entering this opening will find its way out again. Using suffix b to denote a black body, by definition:

$$a_b = 1.$$

Thus, within such a cavity, the radiant flux density at equilibrium is given by:

$$F_b = a_b \mathscr{E}_b = \mathscr{E}_b. \tag{1.1}$$

Consider now the radiation exchanges between a non-black body and the walls of a cavity in which it is situated. At equilibrium the radiant emittance must equal the flux density absorbed, so that:

$$\mathscr{E} = aF_b$$
$$= a\mathscr{E}_b \quad \text{from eqn. (1.1).} \tag{1.2}$$

Thus the radiant emittance of any body at any particular temperature is a fraction of the radiant emittance of a black body at the same temperature. This fraction is the absorption coefficient a of the body at that temperature. This relationship is known as *Kirchhoff's law*.

Since absorption coefficient a can be measured for any surface, $\mathscr{E}$ can be calculated from eqn. (1.2) if $\mathscr{E}_b$ is known. Stefan showed experimentally, and Boltzmann demonstrated on theoretical grounds that:

$$\mathscr{E}_b = \sigma T^4. \tag{1.3}$$

Here T is the absolute temperature in °K (where °K = degrees Kelvin or absolute = °C + 273), and σ is an absolute constant whose value is $5 \cdot 67 \times 10^{-5}$ erg cm^{-2} sec^{-1} deg^{-4}K = $8 \cdot 13 \times 10^{-11}$ cal cm^{-2} min^{-1} deg^{-4}K. Both eqn. (1.3) and the constant σ are associated with the names of Stefan and Boltzmann. From eqns. (1.2) and (1.3), the radiant emittance of any body can be calculated from:

$$\mathscr{E} = a\sigma T^4. \tag{1.4}$$

We turn now to the important question of the distribution with wavelength, or the *spectral distribution*, of radiation. Figures 4 and 5 show the spectral distribution of energy radiated by black bodies at two different temperatures. In both figures the total radiant emittance $\mathscr{E}_b$ is given by the entire area under the curve (i.e. between the curve and abscissa). The ordinates represent what is called the monochromatic radiant emittance, given at any particular wavelength λ by $(d\,\mathscr{E}_b/d\lambda)$. Thus it represents the energy which would be radiated in a wavelength interval 1 cm wide centred on the wavelength concerned, if the flux were constant over this waveband and equal to that for the central wavelength.

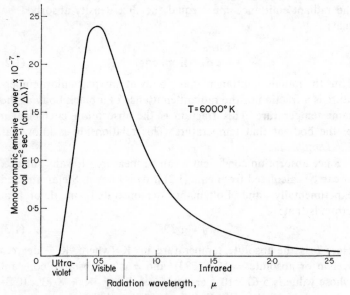

FIG. 4. Monochromatic emissive power versus wavelength λ for a black body emitter at 6000°K, the approximate temperature of the sun.

Measurements show that Fig. 4 approximately describes the wavelength variation of solar radiant flux outside the earth's atmosphere. Thus the sun emits somewhat like a black body at temperature 6000°K; and furthermore the earth emits much as a black body at the relevant surface temperature, in the region of 300°K, which is illustrated in Fig. 5.

Comparing Figs. 4 and 5, it will be noticed that emission reaches a maximum at some particular wavelength (λ_{max}) for any given temperature, and that λ_{max} decreases as temperature increases.

Wien's displacement law describes this feature of the spectral shift with temperature, stating that:

$$\lambda_{max} \propto T^{-1},$$

where the product

$$\lambda_{max} T = 2900, \tag{1.5}$$

when λ is measured in microns (1 μ = 10^{-6} m), and T in °K.

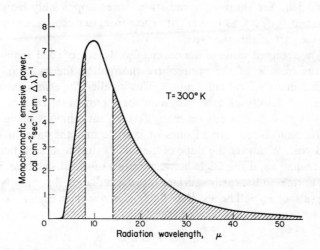

FIG. 5. As for Fig. 4 but at $T = 300°$K, approximate terrestrial temperature. The cross-hatched area under the curve represents radiation absorbed by the earth's atmosphere, and the clear area the waveband where terrestrial radiation is lost to outer space (see text).

Short-wave radiation

Most of the radiant energy from the sun has a wavelength between 0·3 and 3 μ (Fig. 4), roughly half of this energy being radiated with a wavelength in the visible spectrum which is about 0·4 to 0·7 μ.

Comparing Figs. 4 and 5 again, it may be seen that terrestrial radiation is effectively of wavelength $> 3\,\mu$, there being negligible wavelength overlap with solar radiation ($< 3\,\mu$). Terrestrial radiation is thus referred to as *long-wave* radiation to distinguish it from the *short-wave* radiation of the much hotter sun. This distinction is physically important, and useful in discussing radiation exchanges. For example, the reflection coefficient of surfaces for long-wave radiation is very low and approximately independent of the nature of the surface; whereas the reflection coefficient for short-wave radiation varies appreciably between different surfaces, as one might expect from our experience in the visible part of this spectrum.

The accepted value of the normal flux density of solar radiation at the outside of the atmosphere (normal to the radiation) is approximately 2 cal cm^{-2} min^{-1}. This is called the *solar constant*. The flux density of solar radiation for a horizontal area on the earth's surface depends on many factors, the simplest being the astronomical–geometrical ones of latitude and the time of day and year. Whatever the slope of the earth's surface at a particular place may be, if the angle between beams of direct radiation and the normal to the earth's surface is θ, the radiation flux density on the surface F_θ will be less than the flux on a surface normal to the rays F_o, the relation between them being:

$$F_\theta = F_o \cos \theta.$$

Note that at the limit of the earth's atmosphere F_o is equal to the solar constant. Even in the same neighbourhood, θ may vary greatly due to topography, resulting in well known and important differences in microclimate between northerly and southerly aspects of a hill or mountain in non-equatorial latitudes. When soil temperatures are limiting plant growth, advantage can be taken of this same factor by planting on the sun-facing side of cultivation ridges.

Something of the order of 80 per cent of incoming solar radiation can be reflected back into space by a complete cloud cover. Since reflection from clouds occurs at all angles it is referred to as *diffuse reflection*. Solar radiation can also be depleted in its passage through the atmosphere by the processes of *absorption* and *scattering* described in textbooks on optics.

The downward scattered and reflected component of thermal radiation is usually referred to as *sky radiation*. This radiation comes in from all regions of the visible sky, and although it is intense in directions close to that of the sun, it is to be distinguished from the direct beams of radiation. The sum of short-wave radiation directly from the sun and indirectly from the sky is often called the *global radiation*. Its various components are illustrated in Fig. 6.

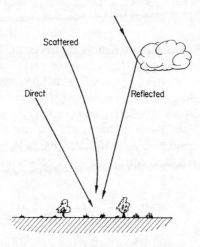

FIG. 6. The various components of global radiation.

Because of the controlling importance of clouds on radiation, fractional cloudiness is commonly estimated visually. Another simple measure of cloudiness is given by the ratio n/N, where n is the total period of "bright" sunshine and N is the maximum period of bright sunshine astronomically possible. The period n is commonly measured with a Campbell–Stokes recorder the essential part of which is an unobstructed glass sphere which collects and focuses radiation on to a calibrated card, producing a burn record when the focused radiation is sufficiently intense. Although the employment of convenient accurate radiation-measuring instruments is rapidly increasing, empirical equations relating incoming short-wave radiation to the fraction n/N will still be of use as they have been in the past. Ångström originally proposed such an equation of the form:

$$\frac{R_S}{F^1} = a + b\frac{n}{N},$$
(1.6)

where R_S is the flux density of incoming short-wave radiation incident upon a horizontal surface at ground level, F^1 is the same as R_S but with cloudless skies, a and b are constants empirically obtained from regression analysis between measured values of R_S/F^1 and n/N.

Albedo is the term used to denote the reflection coefficient for short-wave radiation only. Measurement of the albedo of any ground surface (the term "ground" including vegetation cover) involves suspending an instrument above it which can measure the flux ratio of short-wave radiation across a horizontal surface from below to that from above it (Monteith[11]). An "ideally rough" surface can be conceived for which any incident beam of radiation would be reflected equally in all directions. This is what is

TABLE 1.1. MEASUREMENTS OF ALBEDO NEAR THE EQUATOR
(AFTER E.A.A.F.R.O.)[12]

Type of cover	Albedo (mean daily value)	Remarks
Open water	0·09	Over a sunken evaporation pan 120 cm diam. and 30 cm deep, painted black internally
Bare soil	0·08	Kikuyu Red Loam with a dry dusty tilth.
Short grass	0·21	Closely mown green lawn in dry weather
Bamboo forest	0·12	Mature continuous canopy, 40 ft high
Tea bushes	0·16	Above plucking table of mature bushes
Broad-leaved trees	0·18	Above a canopy 40 ft high

implied by diffuse reflection, and is in complete contrast to the behaviour of light reflected from a mirror. Reflection from natural ground surfaces is of a type somewhere between these two extremes, the intensity of reflected radiation depending both on the angle of incidence of the impinging beam and the direction of observation of reflected energy. Consequently, there is some diurnal variation in albedo with solar altitude, and the albedo of similar surfaces may be different at different latitudes. Some mean daily values of albedo measured near the equator in Kenya are given in Table 1.1.

Long-wave radiation

Total atmospheric absorption and therefore emission of radiation is much greater for long than for short wavelengths. This is chiefly because water vapour absorbs strongly over certain wavelength bands prominent in the terrestrial long-wave radiation spectrum, though carbon dioxide and dust are subsidiary absorbers. This absorption of the earth's outgoing radiation results in a considerable downward long-wave radiation back to the earth from the atmosphere. The *net* long-wave radiation lost to the earth's surface is thus appreciably less than that emitted, and but for this, much lower surface temperatures would occur.

Due to the irregular distribution and intensity of water vapour absorption bands in the infra red, its absorption coefficient varies in a complex manner with radiation wavelength. A crude simplification of this is to say that the absorption coefficient is high for all long wavelengths of terrestrial origin except for a spectral "window" (or region of low absorption) which can be taken to be effectively between 8 μ and 14 μ (Brooks[13] and Sutton[10]). This window straddles the wavelength of maximum emission at terrestrial temperatures, which is about 10 μ for a black body at 300°K (Fig. 5). For wavelengths outside this spectral window, there is always sufficient water vapour in the atmosphere for absorption to be complete. Thus, outside the spectral window, long-wave emission by the ground is at least approximately counterbalanced by radiation back from above. This was indicated by cross-hatching in Fig. 5.

The difference between radiation from the ground surface upwards and the downward long-wave radiation from the atmosphere to ground is known as the *net outgoing radiation* or *nocturnal radiation* from the surface R_L. Since this radiation exchange takes place continuously, the latter term could be misleading. Net outgoing radiation is usually of the same order by day as in the night, but because of the absence of short-wave radiation and the smaller turbulent heat transfer at night this net radiant heat exchange usually dominates the night time microclimate, whilst this may not be so during the day. With an error of less than 10 per cent the ground surface can be regarded as a full or black body radiator, whatever the nature of its coverage, so that if an effective ground surface temperature can be measured (Robinson[14]), outgoing long-wave radiation can be calculated from eqn. (1.3). Thus, provided the downward long-wave radiation from the atmosphere can be estimated at ground level, the net outgoing radiation R_L follows immediately by difference. A knowledge of R_L can be of considerable agricultural and economic importance in areas where frost is possible, as it is one factor determining whether or not frost temperatures will be reached (Sutton, *loc. cit.*, p. 178).

Figure 7 illustrates the components of net nocturnal heat flux, the widths of the arrows representing the magnitudes of the energy fluxes for a possible partition of energy. The downward convection of heat from higher warmer air layers increases with wind, since vertical mixing increases with wind speed; and unless there is an appreciable wind, the net outgoing radiation and the latent heat for any evaporation will be mostly supplied by the heat flux upward from the soil.

R_L can be measured instrumentally as mentioned below, but empirical equations have proved of sufficient accuracy to be very useful if such measurements are not available (Monteith[15] and Swinbank[16]). With clear skies long-wave absorption is almost entirely due to water vapour and carbon dioxide. A much used empirical equation due to Brunt[17] correlates the downward long-wave radiation from clear skies with vapour pressure measured at screen height. Swinbank (*loc. cit.*) has shown that downward long-wave radiation from clear skies can also be closely correlated with temperature measured at screen height.

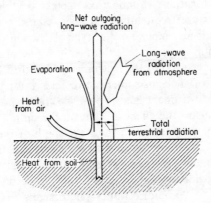

FIG. 7. Components of the *night-time* heat exchange at the earth's surface. (Adapted from Geiger.[8])

An empirical equation for R_L, employing a Brunt-type dependence on vapour pressure (e, mm of mercury), and including the effect of fractional cloudiness c, which has been used by Penman[18] and others in Europe is:

$$R_L = \sigma T_a^4 (0 \cdot 56 - 0 \cdot 09 e^{\frac{1}{2}})(1 - 0 \cdot 9c). \qquad (1.7)$$

In this equation σ is the Stefan–Boltzmann constant, and T_a absolute mean air temperature, both T_a and e being measured in a standard meteorological (Stevenson) screen. In daylight, c can be roughly approximated in terms of actual (n) and possible (N) hours of sunshine by $(1 - n/N)$. It is important to note that the four numerical terms in the two bracketed expressions in eqn. (1.7) can vary appreciably with location, and appropriate values should be determined in the particular location where the equation is to be used.

RADIATION MEASUREMENT

In agricultural research at least two different types of radiation measuring instrument are employed. In connection with plant growth investigations illumination meters or "lightmeters" of various types are frequently used. The exposure meter used in photography is an example of this type of instrument, in which an electrical e.m.f. is generated, frequently by a solid-state device, when light is received by the instrument's detector. In interpreting the readings of such instruments two facts must be recognized. Firstly, each instrument has its own spectral response characteristics. This means the effect of light on the output of the instrument depends very much on the wavelength of the light, just as our own eyes are more sensitive to some wavelengths than others with a maximum sensitivity at about $0.55\,\mu$, and practically no sensitivity outside the waveband 0.4 to $0.7\,\mu$. Secondly, except for specially made "omnidirectional photometers" (Giovanelli[19]), the instrument output should be interpreted in terms of the orientation and solid angle of light acceptance by the light-sensitive receiver of the instrument. The reading of such lightmeters is thus an integration, for light received within a solid angle of a particular orientation, of the interaction between the instrumental factor of spectral response, and the environmental factors of light intensity and wavelength distribution.

Lightmeters are intended to measure illumination, the units of illumination being based on the impression of brightness received by the human eye. It is because the wavebands of visual sensitivity and photochemical effectiveness overlap that illumination meters are used in plant growth investigations. However, because the sensitivity of the human eye varies with wavelength, the relationship between illumination units and units of flux density is not unique, but depends on the wavelength distribution of flux density in the radiation for which the comparison of units is being made. Thus the relationship between illumination and flux density units appropriate for the solar spectrum with a particular state of the atmosphere and for a particular solar altitude is not appropriate under different conditions, and will lead to large errors if applied to the radiation within vegetation or from lamps.

In energy balance studies for example it is necessary to measure energy flux densities in absolute terms, completely ignoring the irregular spectral response characteristics of photochemical reactions or human vision. From the above discussion it follows that illumination meters are inadequate for this purpose.

Absolute flux measurements can be made by absorbing the radiation on a flat surface covered with black paint specially produced to absorb radiation over a very wide waveband. Commonly, this black surface provides the "warm junction" of a thermopile, the "cold junction" of which is a block of metal at air temperature. The thermoelectric e.m.f. produced is then closely proportional to the flux density of radiation absorbed by the black surface, provided the irregular cooling which would be produced by wind is eliminated. Suitable protection can be provided by covering the black surface with a hemisphere of material which absorbs little radiation in the waveband it is desired to measure. Glass is suitable for this purpose if only the short waveband is to be measured. Such an instrument for measuring global radiation R_S (direct plus diffuse short-wave radiation) is called a *solarimeter*. Glass is opaque at long wavelengths, a property utilized in glass-houses, but thin sheet polythene transmits highly in both long and short wavebands. Funk[20, 21] has described a *net radiometer* consisting essentially of both upward and downward facing black plates enclosed in inflated polythene hemispheres which measures the net downward (or upward) radiation for both short and long wavebands. The fundamental importance of net radiation is discussed in the next section. Two such instruments can be used to measure net outgoing radiation R_L (Funk[21]).

Tanner[22] has described in considerable detail the various types of radiation measuring instruments, the principles of their operation and methods of calibration. The subject of calibration is more fully discussed by Morikofer *et al.*[23] and Funk[24].

1.3. Conservation Principles for Heat Energy and Water

The amount and distribution of water as liquid, vapour or solid, and the manner in which the heat energy from the sun is distributed, together have dominance in controlling the day-time microclimate. Application of the principle of energy conservation to the streams of heat energy arriving at, leaving and entering the earth's surface enables an equation of energy balance to be written for the earth's surface, at which the energy conversions occur. Consideration of mass conservation of water similarly yields a water-balance equation. Climate in the layers of air and soil in which plants grow is found to depend strongly on the relative and absolute magnitudes of the various components in these two conservation equations, which provide a physical basis for understanding the infinite variety of microclimates Nature provides.

As we saw in section 1.2 the atmosphere is traversed by a variety of radiation streams which affect meteorological conditions if they are absorbed. Figure 8 illustrates the various components of the day-time heat exchange to a similar scale to that used for night-time exchange (Fig. 7). The relative values of the various energy fluxes can vary enormously, whilst still satisfying the energy conservation requirement. Thus if, as in Fig. 7, the widths of the arrows are taken to be proportional to the magnitudes of the energy fluxes, then the figure shows simply one possible partition of energy.

Considering the proportion of solar energy absorbed by the earth: some is re-radiated with the longer wavelength characteristic of the terrestrial temperature (Fig. 8); some flows into the soil by the process of thermal conduction, thus raising its temperature; absorbed heat can be lost from the ground to the atmosphere by processes of forced and natural convective transport, involving mixing of volumes of air at different temperatures; and, if moisture is present, solar radiation can be used to provide the latent heat necessary for its evaporation, this energy being returned to the atmosphere with the evaporated water. We can express the energy balance in the neighbourhood of the earth's surface by an equation equating all incoming and outgoing energy flux densities:

Net rate of incoming energy per unit area = net rate of outgoing energy per unit area;

$$R_S(1 - \rho) = R_L + G + H + LE \ (\text{cal cm}^{-2} \text{sec}^{-1}), \qquad (1.8)$$

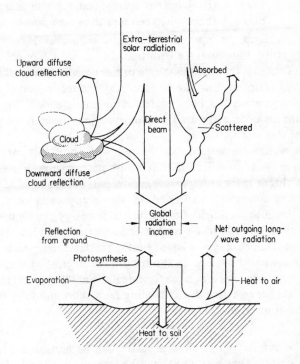

FIG. 8. Components of the *day-time* heat exchange at the earth's surface. (Adapted from Geiger.[8])

where R_S = flux density of total short-wave radiation received by the ground surface from the sun and sky,

ρ = albedo (or reflection coefficient) of the ground surface,

R_L = net flux density of long-wave radiation emitted by

the surface, the difference between that emitted and absorbed,

G = heat flux density into the ground,

H = sensible (or non-latent) heat flux density into the atmosphere,

L = latent heat of vaporization of water (cal g^{-1}), and

E = evaporation rate, negative for condensation (g cm^{-2} sec^{-1}). (Evaporation from plants is also included, though this is often referred to as *transpiration*).

All the terms in eqn. (1.8) have the nature of energy flux density, with units to agree with that chosen for LE of cal cm^{-2} sec^{-1}. Terms on the right-hand side of this equation are positive if energy is removed from the surface. Any change in heat storage of the vegetation has been neglected in this equation, together with the small fraction used in photosynthesis (usually $\sim$ 1 per cent of R_S).

The relative importance of the terms in eqn. (1.8) can vary enormously. If a soil surface is moist or carrying an actively transpiring crop for example, the greater part of the available solar energy is commonly expended in the evaporation of water. However, if the soil dries out so that little energy can be used in this way, there will result a greater loss of sensible heat to the atmosphere. Because it is based on the fundamental principle of energy conservation, eqn. (1.8) must always be satisfied, a change in any one term requiring a readjustment of another term or terms. The total *net radiation flux*, including both short and long wavelengths, is given by:

$$R_N = R_S(1 - \rho) - R_L \qquad (1.9)$$

and this primary climatological factor can be measured directly (section 1.2). Thus eqn. (1.8) may be written:

$$R_N = G + H + LE. \qquad (1.10)$$

We have seen that the amount of moisture present in the soil strongly controls the way in which the net solar energy is utilized. It is, of course, also true that the available solar energy strongly affects the amount of moisture present in the soil. So it is clear that there is a very important interdependence between the energy-balance at the ground surface and the water-balance, which will now be considered.

The water-balance expresses the overall mass conservation for rainfall falling in any given period. Some of it will soak into the soil or remain in puddles on the surface. Some rainfall may flow off the area for which the balance is being studied, and some water may enter by flowing into it from adjacent areas. Also some of the rain received will be evaporated back to the atmosphere. Thus the water-balance equation can be written down by considering the possible fate of the precipitation received (or irrigation water applied) on a certain area of ground in a certain time. The equation is concerned with a volume of soil contained by the imaginary surface which would be generated by a vertical line moving round the area of ground for which the precipitation is considered. The depth of the soil volume considered can be chosen arbitrarily, as will be seen from the equation. Whilst there will be continuous redistribution of moisture (liquid or vapour) between the various terms, and a phase lag as water moves down into the profile, the following conservation or balance equation must be satisfied for any given period and volume of soil.

$$P = S + \varDelta D + \varDelta M + U + \int E \, \mathrm{d}t \; (\text{g cm}^{-2}), \qquad (1.11)$$

where P = precipitation received in the area (of any size) for which the balance is being considered,

$\quad\;\; S$ = net surface run-off,

$\varDelta D$ = increase in surface detention,

$\varDelta M$ = increase in soil water storage due to water flux in any direction across the soil volume considered,

$\quad\; U$ = increase in underground or subsurface storage in layers below that for which $\varDelta M$ was calculated, and

$\quad\;\; t$ = time, so that $\int E \, \mathrm{d}t$ is total evaporation over the period under consideration.

In this equation, evaporation rate E (g cm^{-2} sec^{-1}) is often more conveniently expressed in cm sec^{-1} units, when all the equation's components may be considered as equivalent depth of ponded water (cm). S provides the potential source of surface water supplies, and U of water stored underground. In many situations ΔD may be negligible.

1.4. Conservation Considerations for Air in Contact with the Ground

The possible means of dispersal of radiation and water arriving at the soil surface, and the overall restriction imposed on these means by conservation considerations, were considered in section 1.3. Similar considerations will now be made for a volume of air

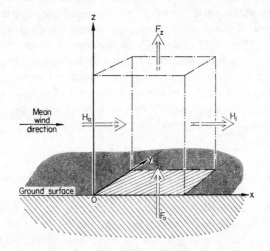

Fig. 9. The broken lines define a rectangular region in space, fixed with respect to the ground surface through which air can flow unhindered. The unit area of ground surface shown in the x–y plane provides the lower boundary to the region.

bounded below by unit area of ground surface, and elsewhere by surfaces perpendicular or parallel to the ground, as shown in Fig. 9. The ground surface is considered uniform, with the mean wind parallel to it and in the direction of the x-coordinate. Denote wind components in the x-, y- and z-directions by u, v and w. It is assumed that mean velocities $\bar{v}$ and $\bar{w}$ are both zero.

Let F_o be the steady upward surface flux of some conservative quantity (such as heat, or vapour with no condensation). (Roughly, the term "conservative" is applied in meteorology to any property of an air mass which remains constant when the air mass undergoes some process. However, whether or not a property is conservative depends on the nature of the process.) The magnitude of this flux at height z (F_z, Fig. 9) would be equal to F_o if the horizontal fluxes of the same quantity, H_0 and H_1 (due to the wind) were also equal. For H_0 and H_1 to be identical, flux F_o must be constant for the ground surface upwind of the element under consideration, and windspeed must be constant. The more important consideration is that if F_o is not constant for a sufficient distance upwind, the gradient in flux H will be of sufficient magnitude that $(H_1 - H_0)$ is not negligible compared with F_o. The quantity $(H_1 - H_0)$ is known as the *advective flux* from the air volume. It is the net result of the transport by the wind of some atmospheric property (such as heat energy or specific humidity) along a gradient of the particular property in question.

Conservation considerations for the rectangular volume of air shown in Fig. 9 yield:

$$F_o = F_z + (H_1 - H_0) + \mathrm{d}C/\mathrm{d}t, \qquad (1.12)$$

where $(\mathrm{d}C/\mathrm{d}t)$ is the change per unit period in the conservative quantity stored within the volume whose vertical and horizontal fluxes are denoted by F and H. The *storage* term C obtained on integrating eqn. (1.12) is very small compared with the total flow per unit area such as $\int F_o \, \mathrm{d}t$. Since the fluxes (of water vapour or heat energy, for example) in all but deep layers of air are large compared with storage changes, the term $(\mathrm{d}C/\mathrm{d}t)$ in eqn. (1.12) can usually be neglected. It then follows that $F_z = F_o$, as is often assumed (at least up to several metres above the ground), only if horizontal advection is negligible. On large open uniform sites, and for periods of the order of a day, net advective effects can be small (Dyer[25]). Even on a uniform site, quite large differences between F_o and F_z due to advection were found over periods of 5 min.

Horizontal advection has another effect besides causing variations in F_z with height. The microclimate characteristic of any one surface is carried downwind, and will result in advection if this crosses a boundary to a surface with different vertical fluxes of moisture or heat. Thus there is a tendency for the above ground changes in microclimate accompanying any sudden contrast in surface type and surface fluxes to be smoothed out. For example, dry air blowing from an extensive arid region across an irrigated area increases in humidity only gradually after crossing an irrigation boundary, with advection of sensible heat. De Vries[26] has given a theoretical analysis of such situations, with application to measurements on irrigated and non-irrigated pastures in the Australian Riverina. Especially with small fields and contrasting moisture regimes, these effects can be considerable.

Bibliography

1. PENMAN, H. L., Weather and crops, *Quart. J. R. Met. Soc.* **88,** 209 (1962).
2. DAVIDSON, J. L., and PHILIP, J. R., Light and pasture growth, *UNESCO Arid Zone Research*, **11,** 181 (1957).
3. LOOMIS, R. S., and WILLIAMS, W. A., Maximum crop productivity: an estimate, *Crop Sci.* **3,** 67 (1963).
4. DE WIT, C. T., Potential photosynthesis of crop surfaces, *Neth. J. Agric. Sci.* **7,** 141 (1959).
5. PENMAN, H. L., Evaporation: an introductory survey. *Neth. J. Agric. Sci.* **4,** 9 (1956).
6. RASCHKE, K., Heat transfer between the plant and the environment, *Ann. Rev. Plant Physiol.* **11,** 111 (1960).
7. PENMAN, H. L., and LONG, I. F., Weather in wheat: an essay in micrometeorology, *Quart. J. R. Met. Soc.* **86,** 16 (1960).
8. GEIGER, R., *The Climate Near the Ground.* Amplified 2nd ed. Harvard University Press, Massachusetts, 1959.
9. VAN WIJK, W. R. (ed.), *Physics of Plant Environment.* North-Holland, Amsterdam, 1963.
10. SUTTON, O. G., *Micrometeorology.* McGraw-Hill, New York, 1953.
11. MONTEITH, J. L., The reflection of short-wave radiation by vegetation, *Quart. J. R. Met. Soc.* **85,** 386 (1959).
12. East African Agricultural and Forestry Research Organization. *Annual Report*, p. 12 (1959).

13. BROOKS, F. A., *An Introduction to Physical Microclimatology*. University of California, Davis, 1960.
14. ROBINSON, G. D., Notes on the measurement and estimation of atmospheric radiation, *Quart. J. R. Met. Soc.* **73**, 127 (1947).
15. MONTEITH, J. L., An empirical method for estimating long-wave radiation exchanges in the British Isles, *Quart. J. R. Met. Soc.* **87**, 171 (1961).
16. SWINBANK, W. C., Long-wave radiation from clear skies, *Quart. J. R. Met. Soc.* **89**, 339 (1963).
17. BRUNT, D., Notes on radiation in the atmosphere, I, *Quart. J. R. Met. Soc.* **58**, 389 (1932).
18. PENMAN, H. L., Natural evaporation from open water, bare soil, and grass, *Proc. Roy. Soc.* Series A, **193**, 120 (1948).
19. GIOVANELLI, R. G., An omnidirectional photometer of small dimensions, *J. Sci. Instrum.* **30**, 326 (1953).
20. FUNK, J. P., Improved polythene-shielded net radiometer, *J. Sci. Instrum.* **36**, 267 (1959).
21. FUNK, J. P., Improvements in polythene-shielded net radiometers, *Proc. Symposium on Engineering Aspects of Environmental Control for Plant Growth* (Melbourne), p. 248 (1963).
22. TANNER, C. B., *Basic Instrumentation and Measurements for Plant Environment and Micrometeorology*. Soils Bulletin No. 6, Dept. Soil Sci., Univ. of Wisconsin, 1963.
23. MORIKOFER, W. (Chm), *et al.*, Radiation instruments and measurements, *Ann. Int. Geophys. Year*, **5**, 365 (1958). (Pergamon Press, N.Y.)
24. FUNK, J. P., A note on the long-wave calibration of convectionally shielded net radiometers, *Arch. Met. Geoph. Biokl.* (B), **11**, 70 (1961).
25. DYER, A. J., Measurements of evaporation and heat transfer in the lower atmosphere by an automatic eddy-correlation technique, *Quart. J. R. Met. Soc.* **87**, 401 (1961).
26. DE VRIES, D. A., The influence of irrigation on the energy balance and climate near the ground, *J. Met.* **16**, 256 (1959).

CHAPTER 2

The Physical Environment of Agriculture: Part II

THE net radiation absorbed R_N might well be regarded as the primary climatological factor, since as mentioned in section 1.3, the microclimate in which plants grow is dominated by the magnitude of R_N, and the components of "sensible" and "latent" heats into which this is partitioned. R_N was defined in section 1.3 as the flux difference between solar and sky radiation absorbed and net long-wave radiation emitted. Symbolically [eqn. (1.9) again]:

$$R_N = R_S(1 - \rho) - R_L. \tag{2.1}$$

This describes the energy input (during daytime) or output (during hours of darkness) as the resultant of all the radiative fluxes to and from the ground surface. Figure 10 gives a typical record of the variation of net and global radiation during a day and a night. At night, with no incoming short-wave radiation, R_N is negative, being equal to the ever-present net long-wave radiation from the ground R_L. At night, R_N is considerably less in magnitude than clear-sky day-time values—which can be of the order of 1 kW m^{-2}. Even before sunset, outgoing radiation usually exceeds incoming, so that R_N is negative and nocturnal cooling has commenced.

The sensible heat flux component of R_N may itself be partitioned into the heat flux into the soil, discussed in section 2.1, and the heat flux into the air in contact with the ground, considered in section 2.4. The theoretical background to some methods of determining the latent heat flux component of R_N is given in section 3.2.

The measurement of microclimatic factors has been discussed in some detail by Slatyer and McIlroy,[1] whilst Geiger[2] provides a comprehensive description of microclimatic factors and their interaction near the ground.

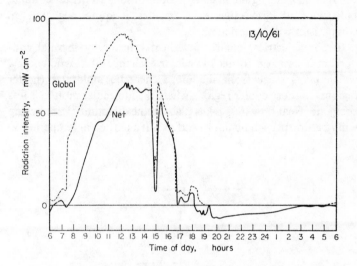

FIG. 10. A record illustrating the variation of global and net radiation (the latter over grass) during mainly fine conditions, with intermittent cloud. (Record taken at Aspendale, Australia. After Slatyer and McIlroy.[1])

2.1. Thermal Conduction, Ground Heat Flux and Soil Temperature

Part of the net radiant input (by day) or output (by night) is respectively dissipated or supplied by thermal conduction in the soil, the term G in the energy balance equation (1.8). The magnitude of G is usually in the range $(0.1–0.3)$ R_N by day (Slatyer and McIlroy[1]), though it is a larger fraction of R_N at night, as is indicated in Fig. 7. The degree of vegetative cover, and, under insolation, the albedo of the ground surface obviously affect the magnitude of G. Conductive transfer can also take place in the air, but except in a very thin layer adjacent to surfaces, convective and radiative transfer are generally so much more effective mechanisms of heat transfer that the contribution due to conduction is negligible in comparison. In soil, on the other hand, radiative and convective heat transfer is quite negligible in comparison with conduction.

In poor electrical conductors, heat transfer by thermal conduction is ascribed to net molecular exchange of kinetic energy, which takes place from the more energetic molecules (hotter regions) to those cooler regions where the molecular motion is less energetic. Heat flow thus takes place in the direction of decreasing temperature, perpendicular to surfaces at a particular temperature.

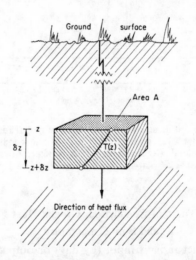

FIG. 11. One-dimensional heat conduction.

If such surfaces are planes, then heat flow is said to be "linear" or one-dimensional, and is illustrated in Fig. 11. The variation of temperature across the element is shown in the figure, and the limit of the temperature drop per unit distance, $\partial T/\partial z$, is known as the temperature gradient. The equation relating rate of conduction of heat Q to temperature gradient is:

$$\frac{dQ}{dt} = -kA\frac{\partial T}{\partial z},$$ (2.2)

where k is the *thermal conductivity* of the material. The negative sign is introduced since heat flow takes place in the direction of decreasing temperature (i.e. $\partial T/\partial z$ negative). In soils k can vary considerably with moisture content, and with soil type. Smith and Byers[3] found k for dry soils to decrease linearly with total porosity (the fraction of any total volume unfilled with solids).

Any difference in heat flux between that entering the upper surface of the elementary volume shown in Fig. 11 and that leaving the lower surface must be used in changing the internal energy and thus the temperature of the element itself. This difference in heat flux will be:

$$-kA\left[\left(\frac{\partial T}{\partial z}\right)_z - \left(\frac{\partial T}{\partial z}\right)_{z+\delta z}\right],$$

where

$$\left(\frac{\partial T}{\partial z}\right)_{z+\delta z} = \left(\frac{\partial T}{\partial z}\right)_z + \frac{\partial^2 T}{\partial z^2}\,\delta z.$$

The rate of increase of internal energy will be the heat capacity of the element ($\rho_b c A\delta z$), where c is the specific heat and ρ_b the density, multiplied by the rate of temperature increase $\partial T/\partial t$. Applying the above energy conservation considerations to the element:

$$kA\frac{\partial^2 T}{\partial z^2}\delta z = \rho_b c A\delta z \cdot \frac{\partial T}{\partial t}.$$

Whence the equation of heat conduction is

$$\frac{\partial T}{\partial t} = \kappa\frac{\partial^2 T}{\partial x^2},$$ (2.3)

where $\kappa = k/\rho_b c$ is called the *thermal diffusivity* or *thermometric conductivity* of the medium. The c.g.s. unit of κ is cm^2 sec^{-1}. In a situation where the temperature of an element of soil is altering, the thermal capacity ($\rho_b c$) of the soil governs the temperature change that will result from the change in quantity of heat in the element, this latter quantity being dictated by thermal conductivity k. Thus the reciprocal of thermal diffusivity, $1/\kappa$ (sec cm^{-2}), will control the time necessary for the soil element to heat up by conduction in any particular situation, the larger the value of $1/\kappa$ ($= \rho_b c/k$), the longer it will take to heat up. On the other hand, a large value of thermal diffusivity will lead to rapid changes in temperature. The soil heat flux component G is continually varying, with daily and seasonal rhythms. The resulting soil temperature distribution then depends on the variation in κ with depth in the soil.

In dry soil κ is of the order of 5×10^{-3} cm^2 sec^{-1}. All the thermal properties of soil depend markedly on moisture content and total porosity. Porosity can be controlled by cultivation, and West[4] found that the diffusivity of a compact soil was reduced to about one-fifth of its original value by cultivation. The lower total porosity of rocks and sands tends to give them greater diffusivities than loams and clays. The thermal conductivity of water

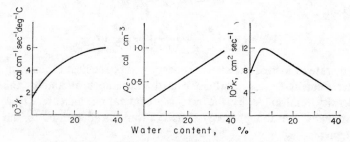

FIG. 12. Generalized curves illustrating the variation of thermal conductivity k, heat capacity ρc and thermal diffusivity κ with water content in soil. (After Rider.[5])

($1\cdot4 \times 10^{-3}$ cal cm^{-1} sec^{-1} deg^{-1} C) is less than that of dry compact soil material. But the conductivity of air (6×10^{-5} cal cm^{-1} sec^{-1} deg^{-1} C) is two orders of magnitude smaller than for water. Thus the replacement of air between soil particles by water materially increases the thermal conductivity of the medium as a whole. The general form of the changes in the thermal properties of soil with moisture content are shown in Fig. 12. The scale values are to be regarded as no more than typical. The decrease in diffusivity of higher moisture contents may explain the general observation (Richards *et al.* in Shaw[6]) that well-drained soils show faster changes in temperature in response to environmental changes than the same soils with higher moisture content.

The specific heat of dry soil is approximately $0 \cdot 2$ cal g^{-1} deg^{-1} C. Shaw,[6] Geiger[2] and Keen[7] (who summarizes much of the early work on the thermal properties and heat flow in soils) all give tables of thermal properties of various soil types. The heat capacity per unit volume of soil, $\rho_b c$, can be calculated usually to within about ± 5 per cent, using a formula due to de Vries[8]:

$$\rho_b c = 0 \cdot 46 \theta_m + 0 \cdot 6 \theta_o + \theta_w, \qquad (2.4)$$

where θ_m, θ_o and θ_w are the volume fractions of mineral material, organic matter, and water respectively, $\rho_b c$ being in units cal cm^{-3} deg^{-1} C. It is to be noted that $(\theta_m + \theta_o + \theta_w + \theta_a) = 1$, where θ_a is the volume fraction of air.

The main difficulty in applying eqn. (2.3) to heat conduction in soils is the variation in κ with depth and time, due to differences in moisture content and porosity, and to moisture movement. Mass movement of moisture completely invalidates the application of eqn. (2.3), since it was assumed in its derivation that heat transfer was by thermal conduction alone. The bodily movement of water through soil can transfer appreciable quantities of heat with it, and in soils with sufficient large pores to allow rapid flow of water this mechanism of heat transfer can swamp that due to thermal conduction. Assuming eqn. (2.3) to be applicable, and calculating an effective value of κ (which could then be regarded only as an "apparent" diffusivity), Callender and McLeod[9] found values varying from $0 \cdot 0016$ cm^2 sec^{-1} under frozen conditions up to a maximum of $0 \cdot 33$ cm^2 sec^{-1}, which is so high that it could only be explained as due to moisture movement. In drier soil conditions moisture can move in the vapour phase under the vapour pressure gradient set up by temperature gradients, also invalidating the application of eqn. (2.3). Because of such effects, the values of experimentally determined thermal conductivity and diffusivity of moist soils may simply depend on the experimental arrangement used. Smith[10] subjected soil samples to a steady temperature gradient for thermal conductivity measurement, and discovered that water movement had taken place from the warm to the cool side of the soil specimen. Gurr, Marshall and Hutton[11] concluded that the apparent equilibrium state of a closed soil column subject to a steady temperature gradient involves a cyclic transfer of moisture, and suggested that moisture moved as a vapour from the warm to the cold side, balanced by a return flow in the liquid phase. De Vries[12] has given the theory on which the results of such experiments could be interpreted. Non-stationary methods of determining thermal constants, such as that described by de Vries,[13] can be sufficiently rapid for moisture transport during measurement to be negligible.

Figure 13 shows the temperature variation over a number of

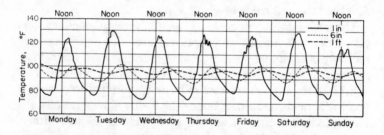

Fig. 13. Typical daily variation in soil temperatures in summer. (January 16–22, 1939, at Griffith, Australia. After West.[14])

summer days at three different depths in a clay soil. The decrease in amplitude and phase delay as the temperature wave penetrates the soil can be seen. The phase delay is illustrated by the time of temperature maximum or minimum becoming later with increasing depth. Figure 13 also illustrates the rapid damping out of irregular short-period fluctuations with depth of penetration, and this is even more noticeable in the annual temperature wave illustrated at two depths in Fig. 14. Despite the important reservation already mentioned concerning the application of eqn. (2.3) to heat flow in soil, the above common *general* features of propagation of the waves of temperature into the soil due to the periodic daily and seasonal fluctuations in insolation can be understood by assuming the soil to be a semi-infinite medium of uniform thermal

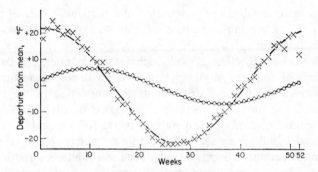

FIG. 14. Record of the annual temperature wave at depths of 1 m ($\times$) and 8 ft (o) with fitted sine curves. (After West,[14] same location as for Fig. 13.)

diffusivity. Let it be further assumed that the temperature of the soil surface is a sinusoidal periodic function of time (a better approximation for the annual than the daily temperature wave), and that it has been so for some time prior to the cycle considered. Then eqn. (2.3) leads to a solution (see Carslaw and Jaeger[15]) in which the amplitude of the wave decreases with depth z below the soil surface in accordance with the function $\exp[-z(\omega/2\kappa)^{\frac{1}{2}}]$, where ω is the angular frequency of the surface temperature. Thus $\omega = 2\pi/\tau$, where τ is the period of surface temperature fluctuations. From this it follows that the amplitude of the temperature wave will have fallen to e^{-3} or 5 per cent of its surface value when $z(\omega/2\kappa)^{\frac{1}{2}} = 3$. Taking $\kappa = 0.005$ cm^2 sec^{-1} as typical, and $\tau = 86{,}400$ sec for the daily wave, z is about 35 cm or 14 in. For the annual wave the corresponding value of z is approximately 6·7 m or 22 ft, illustrating the increase in wave penetration with increase in period.

With the same assumptions, the speed of propagation of the temperature wave is constant and given by:

$$v = (2\kappa\omega)^{\frac{1}{2}}$$
$$= \frac{\Delta z}{\Delta t}, \tag{2.5}$$

where Δz is the depth of penetration of a particular *phase* of the wave (e.g. a maximum) in time interval Δt. For the daily wave, again taking $\kappa = 0\cdot005\,cm^2\,sec^{-1}$, $v = 3\cdot1\,cm\,hr^{-1}$.

Particularly in climates with high insolation rates, the daily pattern of soil and air temperatures is an important environmental factor. In bare dry conditions conduction in the soil controls the surface temperature, the maximum (day-time) and minimum (nocturnal) temperatures being of particular agricultural significance since young plants are susceptible to injury from temperature extremes. Under such soil conditions and with clear skies the maximum daily (downward) heat conduction occurs 2 or 3 hr earlier than the maximum air temperature near the surface which also corresponds to the greatest convectional heat loss to the air (Brooks[18]). Vegetation considerably reduces the temperature range experienced, chiefly because radiation absorption is then distributed through the height of vegetation in contrast to the very thin absorption layer with bare soil. Geiger[2] gives examples of the effect of this on soil temperature and on the temperature distribution in the air. A secondary effect of vegetation can be to alter the quantity of radiation absorbed.

From the viewpoint of energy balance the quantity of direct interest is G, the heat flux in the soil. Deacon[16] has described heat flux plates or meters which, after calibration, can measure this quantity directly in terms of the temperature difference across the plate in equilibrium with the soil in which it is embedded. The possibility of plates impeding moisture movement should be considered. Philip[17] has discussed the theory of such flux meters and given recommendations concerning construction and calibration to reduce possible errors.

Whilst the magnitude of G can be relatively small compared with daytime values of net radiation, its effect in controlling plant microclimate, and on the formation of dew or fog can be of greater importance than might be recognized at first (Rider[5]). But perhaps the greatest agricultural significance of G is its role in influencing soil, and therefore air, temperatures. The increase in molecular kinetic energy with temperature increases the speed of chemical reactions and diffusion processes. Biochemical processes and reactions so fundamental to plant life are partly controlled by temperature-dependent regulating mechanisms. Many physiological processes in plants appear to increase in rate with temperature up to some optimum temperature, and then to drop again, often more rapidly, if such temperatures are exceeded. In particular the rate of respiration whereby chemical energy is made available for the activities of plant cells increases with temperature up to some critical value. Since the rate of photosynthesis of carbohydrates is relatively unaffected by temperature over the range 10 to 25°C, carbohydrate limitation or even depletion can take place at higher temperatures, thus restricting or preventing any increase in dry weight of the plant. The development and activity of microorganisms is also affected by soil temperature (Richards et al. in Shaw,[6] who surveys the effects of soil temperature on plant growth).

2.2. General Features of the Atmosphere near the Ground

An understanding of processes in the lower level of the earth's atmosphere is involved in the study of the above-ground part of the plant's environment, and in elucidating the heat flux terms H and LE (corresponding to the fluxes of sensible and latent heat) in the energy balance described by eqn. (1.8). We shall find that the magnitude of each of these two fluxes is closely associated with the flux of momentum of the air, which in turn depends on air movement and wind structure near the ground.

Especially during the day, the speed and direction of wind is continually fluctuating, and this common observation is confirmed by the records of suitable instruments such as a pressure-tube anemometer. The ultimate origin of atmospheric motion is the unequal heating of the earth's surface by the sun. This inequality in energy input and thus in air temperature is partly compensated by heated air rising above the region of the earth receiving greatest insolation, moving in the upper atmosphere to polar regions, subsiding there, and travelling back nearer the surface of the earth in the opposite direction. This planetary scale *general circulation*, complex though it is (Shaw[19]), is the long-term mean flow pattern of transitory smaller scale circulations associated with weather changes. Up to heights of about 0·5 to 1 km from the earth's surface, air flow is affected by the friction it experiences in moving relative to the earth's surface, and is further affected by air currents associated with local topography or the unequal surface temperatures of land and water. On a microclimatic scale, all this typically results in the gusts and wind fluctuations mentioned above. Thus the atmosphere experiences eddying motions with scales varying from the planetary to the molecular; and it appears that energy is continually passing from the larger to smaller eddies, and is ultimately absorbed by viscosity into the molecular motion which we call heat.

For such reasons the atmosphere is continually being effectively "stirred". Although stirring a tank of water, for example, tends to produce a uniform temperature throughout the tank, this is not true of the atmosphere on a large scale. Above the immediate influence of the earth's surface, air temperature is found to decrease linearly with altitude up to what is called the *tropopause*, by about 0·6°C per 100 m. At least a partial explanation of this decrease can be given (e.g. Sutton[20]) on the basis of several assumptions. One assumption is that the decrease in density or pressure with height is that predicted for an atmosphere at rest. Another assumption is that if a volume of air is displaced from one level to another, no heat exchange with the surrounding atmosphere is necessary for it to be in equilibrium with its new surrounding. This, together with the first mentioned assumption, implies a linear temperature decrease with altitude—referred to as the *adiabatic lapse rate* Γ—which is close to 1°C per 100 m.

The term "lapse" indicates a decrease in temperature with altitude. This is characteristic of the day-time *temperature profile* (the expression for a representation of temperature as a function of height above a surface), because day-time earth-surface temperature usually exceeds that of the air above it. At night the reverse is the case and, near the ground, temperature increases with height, a situation referred to as a *temperature inversion*. Lapse and inversion conditions are illustrated in Fig. 15.

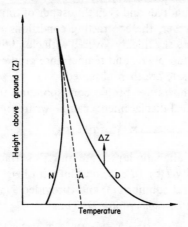

FIG. 15. Illustrating typical day-time lapse D and night-time inversion N temperature profiles. Line A corresponds to the adiabatic lapse rate.

The significance of the adiabatic lapse rate is that it is the temperature gradient for *neutral static stability*, and it provides a criterion for the type of static stability of the atmosphere under other gradients of temperature. Consider temperature profile D in Fig. 15, in which the temperature gradient $\partial T/\partial z$ is (numerically) greater than the adiabatic lapse rate. Any upward displacement of a volume of air in such conditions (indicated by Δz) results in its being at a higher temperature and therefore lower density than its surroundings. There is thus a buoyancy force tending to continue such an ascent, and a temperature profile such as D

indicates that the atmosphere is *statically unstable*. Similar consideration of a vertically displaced volume when $\partial T/\partial z$ is numerically less than the adiabatic lapse (profile N, Fig. 15) shows this type of temperature profile to be the requirement for *static stability*. Though we have not considered stability when the air is in motion, it may be foreseen that this provides a clue to why fluctuations in wind speed (one characteristic of a statically unstable atmosphere) are typically greater by day than by night.

Temperature gradients near the ground under clear sky conditions can be hundreds and even thousands of times the adiabatic lapse rate. However, the contrasting conditions of moderate or high wind under a sky thickly covered with cloud tends to produce a well-stirred atmosphere, and temperature gradients close to the adiabatic lapse rate are then observed.

The linear temperature profile corresponding to neutral equilibrium for vertical displacements can be written:

$$T(z) = T_0 - \Gamma z,$$

where Γ is the adiabatic lapse rate and T_0 surface temperature. Thus the quantity $T(z) + \Gamma z$ is constant with height if the atmosphere is in neutral equilibrium, and (provided $T(z)$ is in °K) it is called the *potential temperature* Θ. Thus:

$$\Theta = T(z) + \Gamma z = T_0,$$

and Θ is the temperature which any volume of air would assume if brought adiabatically to a standard (earth-surface) pressure. An atmosphere is in stable, unstable, or neutral equilibrium according as the potential temperature increases, decreases, or remains constant with height. (Sutton[20] gives a fuller discussion.)

VISCOSITY AND ITS EFFECTS

Suppose a steady flow of air is induced through a long circular pipe by the application of a steady pressure difference between its ends. It will be found on measuring the flow velocity that this is not uniform across the pipe and, if the volume flow rate is not too high, the flow velocity varies in a manner illustrated by Fig. 16a. If a slow uniform air flow could be induced parallel to an extensive flat surface the velocity profile $u(z)$ would be as illustrated in Fig. 16b. Observation shows that the velocity of flow relative to any solid surface tends to zero as the surface is approached, so that the air at the surface can be considered as adhering to it.

The velocity profiles of Fig. 16 suggest that there is a tangential

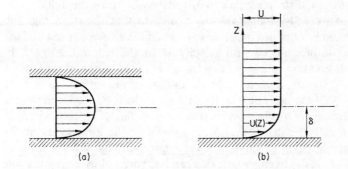

FIG. 16a. Illustrating the *velocity profile* across a steady slow-speed airstream through a long pipe of circular cross-section.

FIG. 16b. Illustrating the velocity profile with uniform steady air flow parallel to an extensive flat surface. Velocity of air flow $u(z)$, more removed from the wall than δ, is constant and equal to U.

shearing stress between the air and the surfaces over which it is flowing, and that this stress is transmitted to some extent to layers of air at some distance from the surface by intermediate layers of air. Such velocity profiles can be explained if we assume that there is an internal friction between layers of air in relative motion, the force per unit area parallel to the *streamlines* being equal to the *shear stress* τ.

Newton (*c.* 1687) made two hypotheses concerning this shear stress, which can be shown indirectly to be valid for many fluids (including air and water), provided the flow is of the streamlined type illustrated in Fig. 16. (This type of flow will be more carefully described in what follows, and designated *laminar* flow.) These hypotheses were that the shear stress between adjacent fluid layers is proportional to the rate of shear at the layer in question; and that the shear stress is independent of fluid pressure. The rate of shearing can be easily shown to be equal to the velocity gradient at right angles to the direction of flow. Thus:

$$\tau = \mu \frac{\partial u}{\partial z}, \tag{2.6}$$

where μ is a property of the fluid, known as the *dynamic viscosity*, a measure of the resistance to distortion offered by the fluid.

An exactly similar velocity profile to that shown in Fig. 16b would be obtained in the steady state if the air were at rest and the flat surface moved with velocity U to the left in the figure. In this situation we can see that the effect of the viscous shear stresses is to transfer something of the motion of the surface up into the bulk of the air not directly in contact with it. Reverting to the situation of air flow over a stationary surface (Fig. 16b), viscosity similarly results in the slowing down of the air velocity at the surface being effective in some degree through a finite layer of thickness δ. In either case this can be expressed by saying that the effect of viscosity is to cause a *diffusion of momentum* through the

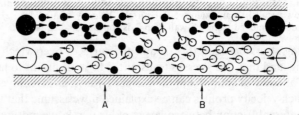

FIG. 17. Plan view at a particular instant of two streams of people moving in opposite directions. Arrows indicate direction of motion.

air in a direction normal to the flow. It is due to one layer of air tending to move adjacent layers with it because of the shear stress between them.

But what is the origin of these shear stresses? It is a consequence of the molecular structure of the fluid and may be more readily understood in terms of the following analogy. Figure 17 represents a view from above of two crowded streams of-people rushing in opposite directions. They are kept separate by a dividing wall, except between A and B. If the streams are closely packed and really in a hurry, the figure illustrates the possible fate of individuals. A person forced from either stream across into the other carries with him his momentum characteristic of the stream from which he has come. In this situation this causes a decrease in the momentum of the stream entered. This transfer of momentum from one region to another where the "bodies" have a different velocity is effectively a shearing stress between the two streams. It is a consequence of Newton's laws of motion that the shearing stress between two layers is equal to the net transfer of momentum per second across unit area dividing the two layers; and that the shearing stress is equal but oppositely directed on either layer. These considerations would be equally true if both streams were travelling in the *same* direction, though there would be a shear stress only if their velocities were different.

Whilst the analogy is basically interpreted if the people are considered to be fluid molecules, the analogy is deficient in one particular. This is that fluid molecules possess an incessant thermal motion in random directions, and in gases the distance travelled by a molecule between successive collisions is about 1000 times greater than molecular dimensions. Thus a much more effective interpenetration between layers by molecules takes place than is hinted at by Fig. 17. The kinetic theory understanding of viscosity is that in laminar flow with a velocity gradient—whether the gradient is due to proximity to a solid surface or for any other reason—the random thermal motion of the molecules results in a net transfer of momentum from regions of higher bulk velocity to regions where this is lower. The "bulk velocity" refers to the over-all drift velocity—a reflection of mean net molecular motion over time intervals much greater than those between successive col-lisions of a molecule. If such ideas are worked out quantitatively for gases (as is done for example by Sutton[21]), the viscosity coefficient is found to be related to molecular properties by:

$$\mu = 1/3\rho_a cl, \tag{2.7}$$

where ρ_a is the gas density, c the mean velocity of gas molecules, and l the mean distance between successive collisions (the *mean free path*).

The viscosity of air at 20°C is only $1 \cdot 8 \times 10^{-4} \mathrm{g\,cm^{-1}\,sec^{-1}}$. Thus from eqn. (2.6) velocity gradients must be considerable before viscous shear stresses become appreciable. Such high gradients are found only in proximity to solid surfaces. It is a fact of great significance that the velocity gradient in the situation illustrated by Fig. 16b, for example, decreases extremely rapidly with distance from the surface. Thus, as Prandtl first made clear in 1904, viscosity has virtually no direct influence on flow outside a very thin layer of air (usually only of the order of a millimetre thick for limited surfaces) adjacent to any solid surface. Prandtl introduced the term *boundary layer* to describe this concept—a concept of fundamental importance in all problems concerned with the motion of air and heat transfer across solid–fluid interfaces. Sutton[22] gives a clear and simple discussion of the boundary layer and other factors of importance in understanding the resistance experienced due to relative movement of solid bodies and air.

We have so far been concentrating on the transport of momentum due to the molecular mixing process. The same process could also be regarded as a transport of molecular kinetic energy or of mass, thus giving a kinetic interpretation of thermal conduction and diffusion in fluids. But such molecular processes are so slow in action that some much more effective mechanism than molecular transport must be responsible for the efficient mixing in the lower regions of the earth's atmosphere, where both heat and mass transfer can take place at rates many orders of magnitude greater than would be possible with molecular processes alone. The reasons for this will now be given.

LAMINAR AND TURBULENT FLOW

It was the now well-known experiments of Osborne Reynolds the flow of water in pipes which first called scientific attention to the two basically different forms in which fluid flow can occur. Flow is described as *laminar* if adjacent volumes of fluid move in a regular manner with respect to each other. In *steady* flow streamlines give the direction of motion of a given fluid element, so that in steady laminar flow streamlines will remain fixed in space, as is illustrated in Fig. 16. In *turbulent* (or *eddying*) flow on the other hand the velocities of fluid elements vary in an irregular and apparently random manner. If such flow is investigated with an instrument which can follow very rapid changes in velocity of the air, the fluctuation in velocity at a given point and in a given direction may appear as in Fig. 18. The velocity component at

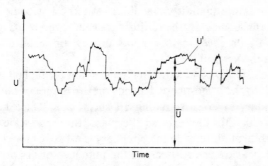

FIG. 18. Fluctuations in wind velocity in a given direction (u).

any instant can be regarded as the sum of a *mean velocity* $\bar{u}$ and a fluctuation or *eddy velocity* u' such that the mean value of u' over some specified time interval is zero.

Any flow in the atmosphere near the ground, with the possible exception of short periods with very calm conditions at night, is characteristically turbulent. This fact dominates the spatial distribution of wind speed, water vapour and temperature, and together with the energy and water balances, controls the magnitudes of the fluxes of momentum, water vapour and heat. Friction between earth and atmosphere, together with convectional plumes of air are causes of atmospheric turbulence, as will be discussed in more detail later.

In turbulent flow close to the ground surface, fluctuations in the path of volumes of air are of a scale comparable with that of surface irregularities (Sutton[20]). In contrast the rate of diffusion of motion due to viscosity is limited by the exceedingly small distance between molecular collisions (about 10^{-5} cm in air near the ground), even though something like 10^9 collisions per second are experienced by each molecule. The vastly increased scale of mixing for turbulent air in comparison with that due to molecular heat motion results in mixing and momentum transfer increasing, usually by several orders of magnitude. If this were not so, the air layers inhabited by plants and humans would experience such severe daily extremes of all microclimatic properties, such as temperature and humidity, that life as we know it could not go on. Air turbulence is the diffusing agency moderating these extremes by its effect in thoroughly mixing air in all directions including the vertical.

Heat transfer by conduction and convection from plant and soil surfaces into the atmosphere depends on the nature of air flow in the boundary layer surrounding all such parts. Once heat has traversed this thin film around the plant, the large-scale mixing in the turbulent air outside the boundary is a vastly more effective means of heat transport. Because the influence of viscosity is so much greater in the boundary layer surrounding any object, the nature of air flow in such layers is different in character to that outside it. It is found that even though the bulk air flow near the earth is turbulent, the flow in the boundary layers over fixed solid surfaces with dimensions of the order of plant leaves is normally laminar (Sutton[20]). This is illustrated in Fig. 19 for flow whose mean direction is parallel to a flat solid surface of sufficient extent for *transition* to occur, the scale of turbulence in the turbulent boundary layer being rather less than in the bulk air. The region L denotes a very thin *laminar sublayer*, in which any vertical eddy motion is practically non-existent, even under a turbulent boundary layer. The boundaries in Fig. 19 are dashed as a reminder of their indefinite character.

Within a laminar boundary layer diffusion rates are much

reduced compared with those in the air outside, so that gradients of microclimatic variables are very great. The value of such variables *at* any surface are therefore extremely difficult to measure instrumentally.

The vastly increased mixing in turbulent flow compared with that in laminar flow means the shearing stresses due to turbulent eddy transfer of momentum, called *eddy shearing stresses*, are usually so much greater than purely viscous stresses that the

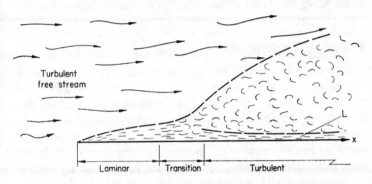

FIG. 19. Laminar boundary layer, undergoing transition to turbulent condition. (Vertical scale greatly magnified in comparison with horizontal.)

latter can be ignored. With regard to the mechanism of net momentum transfer in turbulent flow which takes the place of the molecular transport in laminar flow, consider a steady but turbulent air flow along the ground surface, maintained against the retarding effect of friction at the ground surface. The mean velocity will characteristically decrease as the surface is approached. To maintain this steady state, faster-moving air must be continually moving to lower levels due to turbulent fluctuations, with a simultaneous transfer of retarded volumes of air from below to above. Analysis of turbulent motion (Sutton[20, 21] and section 2.3) shows this process of momentum transfer to depend on a correlation between horizontal and vertical eddy velocities such as u' in Fig. 18.

As is shown in Fig. 19, the thickness of a boundary layer increases with distance downstream from where it began to form. Considering air flow over the earth's surface this distance is effectively infinite, and such flow may be regarded as a "fully developed" boundary layer. At heights in excess of about 2000 ft the effects of the frictional drag of the earth's surface are no longer appreciable, and the rapid small scale fluctuations characteristic of turbulence close to the earth are absent. Such a height could be regarded as the "thickness" of the earth's boundary layer.

FORCED AND CONVECTIVE TURBULENCE

Friction between the ground and moving air is not due to viscosity or to eddy shearing stresses, or is only negligibly so. The reason for it is the same as for the resistance experienced by a cyclist. The flow around any *bluff* (or non-streamlined) body is such that the air pressure is greater in front than behind the body, resulting in a force opposing the motion. This is an example of a *drag* force, and is referred to as *form* (or pressure) drag, since its magnitude depends on the form or shape of the body. The ground surface is effectively a closely spaced array of bluff bodies, and for such a surface the frictional stress is found to be closely proportional to the square of the mean wind velocity at some arbitrary reference height above it. This is not true for a flexible crop surface, or long grass, owing to the change in nature of such surfaces as wind speed increases. The characteristic feature of flow around a bluff body is an extensive disorderly *wake* on its down-wind side, due to detachment of the boundary layer from the bodies' surface.

This ground–air friction is always a cause of atmospheric turbulence. If wind speeds are at least moderate and the sky covered with cloud, temperature gradients between ground and air and in the air itself are low, and approach the adiabatic lapse rate. The atmosphere is thus very close to a condition of neutral static stability, which exhibits behaviour simpler to analyse than that to be described in the next paragraph. Under these conditions atmospheric turbulence is almost entirely due to friction with the ground, and is referred to as *forced*, *mechanical* or *frictional turbulence*.

Under other weather conditions, temperatures at the ground surface exceed those in the air, and temperature gradients numerically greater than the adiabatic lapse rate are normal (see Fig. 15). As was shown earlier, the atmosphere is then statically unstable, and this can result in regions of ascent of air hotter than their surroundings. Various models have been suggested to describe the structure of such convective motions including "bubbles" and "plumes" pulled along by the wind. This instability is associated with *free* or thermal convection. It is an additional cause of turbulence, referred to as *convective turbulence*. The relative importance of convective as against forced turbulence increases with the excess of temperature gradient over the adiabatic lapse value, and decreases with the magnitude of wind speed gradients (Richardson[23]). Static stability of the lower atmosphere which accompanies nocturnal temperature inversions (Fig. 15) tends to reduce or damp out turbulence.

Gradients of microclimatic variables decrease with increase in turbulence, due to the increased mixing. Nevertheless these gradients can be measured by suitably spaced instruments, and such basic information, combined with turbulent transport theory, makes possible a calculation of vertical atmospheric fluxes, as will be described in later sections of this chapter.

2.3. Turbulent Transfer

The limited aim of this section, of those which follow it in this chapter, and of section 3.2, is to give sufficient physical background so that the research literature on transport processes near the ground, including evaporation and heat loss, may be followed. With the degree of turbulence characteristic of day-time conditions, the transports of momentum, heat, and water vapour are all intimately related since each flux is dependent on the turbulent structure of the surface air. This is not completely true since heat exchange by thermal radiation (discussed in section 1.2) is a process not directly influenced by turbulence. At night, unless there are at least moderate winds and considerable cloud, heat fluxes are largely radiative in character. Thus the close association between fluxes of momentum, heat and water vapour characteristic of day-time conditions is by no means universal at night.

The chief reason for this common disparity between day- and night-time flux associations is the different character of air flow which is typical of the night. At night with low wind, statically stable conditions prevail due to air temperature increasing with height near the ground (section 2.2). Turbulence in the air thus tends to subside into calmer conditions. Also at night the lower, cooler, and more dense air layers then tend to flow downhill and into valleys under the action of gravity, and an understanding of these nocturnal density currents is important in frost prediction and protection (Brooks[18]).

In the remainder of this section it is assumed that air flow is *fully turbulent*. The implication of this term is that transport of atmospheric properties due to the bulk mixing of air characterizing turbulence completely swamps transport due to molecular heat motion. This is usually the case in air flow near the ground, though within vegetation under calm nocturnal inversion conditions, molecular transport in the bulk air may not be negligible compared with turbulent transfer. Within the laminar boundary layers normally surrounding plant parts (section 2.2), transport is primarily due to molecular motion. Under simple and well-defined surface and flow conditions, often encountered in engineering problems, transport of heat or matter through such boundary layers can usually be adequately predicted (Eckert and Drake[24]). However with the enormous variety and complexity of surfaces in agriculture across which transport of heat and water vapour is taking place, a rather different approach is adopted for most problems.

It will also be assumed in this section that horizontal advection is negligible, and radiative heat exchange is not considered.

FLUCTUATION THEORY

A general expression for the turbulent flux of any transferable conservative property (defined in section 1.4), per unit mass of fluid c, can be obtained as follows: Since at any point in the earth's atmosphere the vertically upwards velocity is w, the instantaneous upward flux is then $\rho_a w c$, where ρ_a is the density of moist air, the dimensions of the flux depending on the nature of quantity c. The quantity $\rho_a w$ can be measured directly with a suitable anemometer (Taylor[25]). In terms of the vertical eddy velocity w', and corresponding eddy fluctuation c' in quantity c:

Instantaneous flux $\qquad \rho_a w c = [\overline{\rho_a w} + (\rho_a w)'](c + c')$, $\qquad$ (2.8)

where a bar denotes an average over a period of observation, and dashes denote departures from their respective mean values. At a height of the order of one to several metres above an agriculturally uniform surface, $\overline{\rho_a w}$ has been found to be zero over a sufficiently long period of observation (of the order of hours, Dyer[26]). Very close to rough ground or irregular vegetation there can be a long-term mean vertical air motion, so that $\overline{\rho_a w} \neq 0$. However, taking $\overline{\rho_a w} = 0$ in eqn. (2.8):

Instantaneous flux $\qquad \rho_a w c = (\rho_a w)'\bar{c} + (\rho_a w)'c'$. $\qquad$ (2.9)

Now consider the average flux over an "adequate" period of time. The term $(\rho_a w)'$ is eddy momentum per unit volume, and provided conditions are such that $\overline{\rho_a w} = 0$, it follows by definition that $(\overline{\rho_a w})' = 0$. Furthermore, since by definition there can be no correlation between the eddy quantity $(\rho_a w)'$ and the time mean value $\bar{c}$ of the specific conservative quantity c, the average value over an adequate period of the term $(\rho_a w)'\bar{c}$ in eqn. (2.9) must be zero. But the average value of $(\rho_a w)'c'$ will *not* be zero if there is any correlation between $(\rho_a w)'$ and c', and $(\rho_a w)'c'$ is referred to as the *eddy flux*. How such a correlation can arise can be seen by considering some particular examples: For a steady horizontal mean wind near the ground, the velocity u in the x-direction (see Fig. 9) is a conservative quantity. Thus we may firstly take c to be u and c' to be u', the eddy velocity in the x-direction. The steady wind can be maintained against the effect of friction at the ground surface only by faster-moving air from above being brought nearer the surface (together with the simultaneous transfer of retarded air in the upwards direction). This transfer in both directions is affected by turbulence, and there is an analogy (which must not be taken too far) with the transfer of the retarding effect of a surface in laminar flow by the momentum transfer of molecular interchange, as explained in section 2.2. In turbulent (as in laminar) flow the downward rate of momentum transfer per unit area is equal to the shear stress at the surface over which the flow takes place. For this downward transfer of momentum to take place in turbulent flow, gusts (positive u') must be associated more frequently with descending flow (negative w') than with upward currents (positive w'). Thus we would not expect $\overline{w'u'}$ to vanish on averaging over any period. The average $\overline{w'u'}$ is known as the *eddy shearing stress*, and in fully turbulent flow it is very much greater than the viscous shearing stress described in section 2.2.

As a second example consider c to be heat content per unit mass. An eddy with greater heat content than its surroundings will tend to rise because of its lower density, thus resulting in a correlation of w' and c'. Also if c is specific humidity q (gram water vapour per gram of moist air) it is the correlation in eddy components which results in turbulent transport of water vapour.

Thus in general the mean flux $\overline{F}$ per unit area of any specific conservative quantity c due to turbulent transport is given by:

$$\overline{F} = \overline{\rho_a w c} = \overline{(\rho_a w)' c'}. \tag{2.10}$$

Eddy correlation instruments employing eqn. (2.10) have been very satisfactory for sensible heat flux measurement. However some instrumental limitations have yet to be overcome for accuracy in water vapour flux determinations in all situations. A general description of the instrumental arrangements required was given by Taylor and Dyer,[27] and Dyer (*loc. cit.*) has given some examples of measurements obtained.

EXCHANGE COEFFICIENT (OR TRANSPORT CONSTANT) HYPOTHESIS

This hypothesis is that the mean vertical flux of any conservative quantity c is proportional to the gradient of the mean value $\bar{c}$ of the quantity in the vertical direction. The factor of proportionality between the flux and the gradient is called an *exchange-coefficient A*, so that:

$$\text{Mean flux } \overline{F} = A\frac{d\bar{c}}{dz}$$

$$= \overline{(\rho_a w)' \, c'} \quad \text{from eqn. (2.10).}$$

English language writers have mostly followed G. I. Taylor in not using A, but employing K_t defined by $K_t = A/\rho_a$, where ρ_a is the density of (moist) air. K_t is often referred to as a *transport constant*, or *transfer coefficient*.

MOMENTUM TRANSPORT

Consider the continuous downward flux of momentum by turbulence. The appropriate transport constant for momentum K_m is variously referred to as the "coefficient of eddy diffusivity", "eddy transfer coefficient", and, by analogy with eqn. (2.6) describing molecular momentum transport in laminar flow, the "eddy viscosity". K_m has the same dimensions L^2T^{-1} as thermal diffusivity, explaining why it is also referred to as an "eddy diffusivity". From the definition of K_m, it follows that:

Mean flux of momentum per unit area = shear stress τ on the surface across which it is transferred

$$= \rho_a K_m \frac{d\bar{u}}{dz}.$$

Or
$$\frac{\tau}{\rho_a} = K_m \frac{d\bar{u}}{dz}. \tag{2.11}$$

If flow were not fully turbulent then $v \, (= \mu/\rho_a)$—the kinematic viscosity of the fluid—would not be negligible in comparison with

K_m, which should then be replaced in eqn. (2.11) by $(K_m + v)$. This is a significant modification only under extremely calm (near laminar) conditions, or in the boundary layer of a surface, where turbulent transport is suppressed.

CONSTANCY OF MEAN SHEARING STRESS

In section 2.2, and in the above discussion of fluctuation theory, we have seen how momentum is transferred down through the air in turbulent flow, and that the momentum flux per unit area is equal to the shear stress in the plane across which the momentum is transferred. Consider steady but turbulent air flow parallel to a horizontal ground surface. The forces acting on an elementary volume in such flow are shown in Fig. 20, where $\bar{p}$ represents the mean air pressure, and τ mean shear stress. Since the flow is assumed steady, inertia forces are neglected, and the shear stress and pressure forces shown in the figure must be in equilibrium in a horizontal direction.

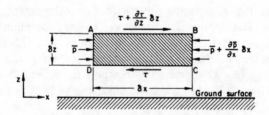

FIG. 20. An elementary region (of thickness δy normal to the figure), fixed with respect to the ground surface.

The shear *force* acting on the horizontal surface CD of area $\delta x \delta y$ will be $\tau \delta x \delta y$, with a similar expression for the oppositely directed shear force on surface AB (Fig. 20). Thus the resultant shear force, in direction AB, will be

$$\frac{\partial \tau}{\partial z} \delta z \delta x \delta y.$$

Equating this to the net pressure force in the opposite direction we get:

$$\frac{\partial \tau}{\partial z} = \frac{\partial \bar{p}}{\partial x}.$$

If pressure gradient $\partial \bar{p}/\partial x$ is supposed independent of z—a reasonable assumption over a limited height—then integration of this equation gives:

$$\tau = \tau_0 + z \frac{\partial \bar{p}}{\partial x}, \tag{2.12}$$

where τ_0 is the value of τ at the surface ($z \to 0$). In the atmosphere, horizontal pressure gradients will rarely exceed the order of 1 mb per 100 km, as may be seen from the spacing of isobars on weather charts. Now τ_0 can be measured directly, essentially by measuring the force on a small representative portion of the earth's surface placed in a shallow pan and floated in oil. For level pasture Pasquill[28] found τ_0 to be of the order of 1 dyne cm^{-2}.

We can now compare the relative magnitude of terms in eqn. (2.12). Since 1 mb = 10^3 dyne cm^{-2}, $\partial \bar{p}/\partial x$ can be of the order of 10^{-2} dyne cm^{-2} m^{-1}. Thus $z(\partial \bar{p}/\partial x)$ will be comparable with τ_0 when z is 100 m (or greater for lower pressure gradients). Thus, with the assumptions given, the approximation

$$\tau = \tau_0 = \text{constant} \tag{2.13}$$

should be reasonable within perhaps the first 10 to 20 m from the ground, and is commonly made in agricultural micrometeorology up to several times crop height. A constant mean shear stress τ implies constancy of downward momentum flux per unit area through this shallow layer, and, from eqn. (2.11), a constancy of the product $K_m(\mathrm{d}\bar{u}/\mathrm{d}z)$. We shall return to the latter implication when considering mean wind-velocity profiles.

TURBULENT HEAT AND WATER VAPOUR TRANSPORT

In discussing viscosity in section 2.2 we saw that the same molecular mixing process gave rise to transport of momentum, heat and matter. Although in turbulent flow molecular mixing is completely overshadowed, the mixing which accompanies turbulence in a similar way must result not only in momentum transfer, but also in the transport of sensible heat and water vapour. However, in contrast to the situation in laminar flow, the mechanisms resulting in the transport of these three quantities in turbulent flow are not necessarily precisely the same, though the differences are not fully understood. Also the boundary conditions, as the earth's surface approached, are not the same for the three fluxes.

It follows from the exchange coefficient hypothesis that we can formally write transport equations for sensible heat flux H (cal cm^{-2} sec^{-1}) and flux of water vapour E (g cm^{-2} sec^{-1}) as:

$$H = -\rho_a c_p K_h \frac{\partial \bar{T}}{\partial z} \tag{2.14}$$

and

$$E = -\rho_a K_w \frac{\partial \bar{q}}{\partial z}$$

$$= -K_w \frac{\partial \bar{\rho}_v}{\partial z}, \tag{2.15}$$

where ρ_a = density of moist air (g cm^{-3}),

c_p = specific heat of moist air at constant pressure (cal g^{-1} deg^{-1}C),

K_h and K_w are eddy transfer coefficients for heat and water vapour respectively (cm^2 sec^{-1}),

q = specific humidity (gram vapour per gram moist air), and

ρ_v = vapour density (or absolute humidity) (g cm^{-3})

= $\rho_a q$.

It is important to realize that the transfer coefficients can vary over an extremely wide range. The magnitude of such a transfer coefficient in a high wind a few metres above a rough surface could be greater than that in a low wind close to a smooth surface by a factor of 10^4 or more (Slatyer and McIlroy[1]). Furthermore, the formal transport equations we have given, which define the transfer coefficients, in themselves give no information concerning the magnitude or manner of variation of these coefficients.

The analogy between turbulent transport of momentum, heat and matter, if assumed exact (as first suggested by Reynolds) implies identity of all three transfer coefficients. This is referred to as *Reynolds' analogy*. Measurements over open uniform sites

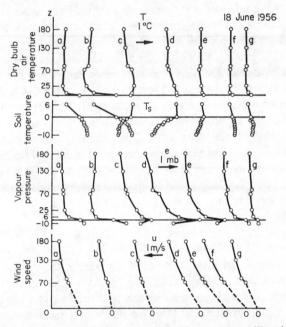

FIG. 21. Profiles of air temperature T, soil temperature (T_s, with z-scale enlarged 5 times), vapour pressure e, and wind speed u at (a) 00, (b) 04, (c) 08, (d) 12, (e) 16, (f) 20, (g) 24 hr, 18 June 1956. Wheat crop, height 70 cm, Rothamsted Experimental Station (after Penman and Long[26]).

(Pasquill,[29] Swinbank[30] and Rider,[31] for example) have tended to confirm equality of coefficients under conditions of neutral atmospheric stability (defined in section 2.2), provided mean values of quantities averaged over periods of about an hour or more are used. However, there are difficulties and uncertainties under other stability conditions. If we take

$$K_m = K_h = K_w = K_t, \tag{2.16}$$

it follows from transport equations (2.11), (2.14) and (2.15) with the further assumptions of eqn. (2.13) and of no horizontal advection, that mean profiles of wind speed, air temperature, and humidity should all be geometrically similar. For consideration in connection with this conclusion, Fig. 21 presents temperature and vapour pressure profiles within and above a wheat crop, together with wind profiles above the crop, at 4 hr periods throughout a day. The curves are displaced relative to each other for clarity, and it should be noted that increasing wind speed is to be left, where for other quantities it is to the right. A study of the daily variation in profiles, and the interrelation between profiles of different properties is profitable. Penman and Long[32] considered the exaggerated lapse in vapour pressure for curve d near $z = 70$ cm as improbable, and the inversions in curves f and g impossible, and showed that they could be due to technical difficulties raised by temperature fluctuations.

MEAN WIND PROFILES NEAR THE GROUND

In cases of agricultural interest the surface of the earth is almost always *aerodynamically rough*, implying that the shear stress τ_o at the surface is almost entirely due to form drag (see section 2.2) of the rough surface, and that viscous shear stresses are negligible in comparison. Consequently, τ_o has the nature of an eddy shear stress. From an aerodynamic point of view the earth's surface is a dense array of "bluff bodies" (see section 2.2), across which flow is a series of turbulent wakes. In turbulent flow shear stresses are found to be approximately proportional to the square of mean relative velocity. Consequently surface shear stress τ_o is closely proportional to the square of a reference velocity taken at some arbitrary height, often 1–2 m above the earth's surface. It is useful to introduce a related velocity, known as the *friction velocity* u_*, for which this square law holds exactly, and not just closely. This we define:

$$u_*{}^2 = \frac{\tau_0}{\rho_a}$$

$$= \frac{\tau}{\rho_a} \quad \text{from eqn. (2.13)}$$

$$= \overline{(u'w')} \quad \text{from eqn. (2.10),} \quad (2.17)$$

with $c' = u'$ and $\overline{F}$ = momentum flux per unit area = τ. (It can be checked that τ/ρ_a has the dimensions of a velocity squared.) From the expression of $u_*{}^2$ in terms of eddy velocities, it follows that u_* must be of the same order of magnitude as u' and w'. Since for rough surfaces τ_0 is closely proportional to $\bar{u}^2$ (at some reference height) it follows that u_* will be nearly proportional to $\bar{u}$, and for grass about 50 cm high $u_* \simeq \bar{u}/10$ for a reference height of 2 m. As would be expected, $u_*/\bar{u}$ increases with surface roughness, ranging from about 3 per cent for very smooth surfaces (such as mud flats) up to about 13 per cent. Typical values of u_* for natural surfaces are given below in Table 2.1. Furthermore, $u_*/\bar{u}$ increases as the surface is approached, showing that the eddy velocities become an increasingly bigger fraction of the mean velocity with proximity to the surface.

The manner in which mean wind speed $\bar{u}$ increases near the ground can be conveniently expressed in terms of its non-dimensional ratio with u_*. The simplest agricultural surfaces from the point of view of wind profiles are those which are covered with stiff short vegetation (less than a few centimetres in height). Postponing the complications with taller flexible crops, measurements over level surfaces of considerable extent and good uniformity of surface roughness (for example, Deacon[33]) have shown that the mean wind profile near the ground under conditions near neutral stability can be represented by:

$$\frac{\bar{u}}{u_*} = \frac{1}{k} \ln \frac{z}{z_0}, \quad (2.18)$$

where k is a constant of proportionality (introduced in turbulent flow problems by von Kármán) having a value close to 0·4. (Since $\bar{u} \to \infty$ as $z \to \infty$, eqn. (2.18) is clearly valid only for restricted heights.) Height z is expressed as a non-dimensional ratio with z_0, called the *roughness length*. From eqn. (2.18) we would formally regard z_0 as the value of z for which $\bar{u} = 0$. However in practice there is departure from this logarithmic type profile at values of z comparable with z_0, and z_0 is related to the intercept on the $\bar{u}$-axis obtained by extrapolation of the plot of $\bar{u}$ versus ln z. z_0 is usually of the order of one-tenth the average height of surface protuberances. Table 2.1 gives representative values of z_0 and u_*.

TABLE 2.1. REPRESENTATIVE VALUES OF z_0 AND u_* FOR NATURAL SURFACES
(Neutral stability; values of u_* corresponding to $\bar{u} = 500$ cm sec^{-1}
at 2 m height)   (After Sutton[20])

Type of surface	z_0 cm	u_* cm sec^{-1}
Very smooth (mud flats, ice)	0·001	16
Lawn, grass up to 1 cm high	0·1	26
Downland, thin grass up to 10 cm high	0·7	36
Thick grass, up to 10 cm high	2·3	45
Thin grass, up to 30 cm high	5	55
Thick grass, up to 50 cm high	9	63

For taller uniform vegetation, but otherwise with the same restrictions as were made concerning eqn. (2.18), mean velocity profiles can be represented by assuming turbulent exchange to commence at some height d, known as the *zero-plane displacement*, above the earth's surface, thus replacing z in eqn. (2.18) by $(z - d)$.

EVALUATION OF TRANSFER COEFFICIENT K_m

If conditions are such that the neutral wind profile of eqn. (2.18) is applicable, conclusions can be drawn concerning the manner of variation of the eddy transfer coefficient K_m, as follows:

Differentiating eqn. (2.18) yields:

$$\frac{\mathrm{d}\bar{u}}{\mathrm{d}z} = \frac{u_*}{k}\frac{1}{z}. \tag{2.19}$$

But from eqn. (2.11)

$$\frac{\mathrm{d}\bar{u}}{\mathrm{d}z} = \frac{\tau}{\rho_a K_m}$$

Thus

$$K_m = \frac{\tau k z}{\rho_a u_*}$$

In the layer for which τ can be regarded constant (and the logarithmic velocity profile does not apply for heights greater than those limited by this restriction), we can substitute $\rho_a u_*^2$ for τ (from eqn. (2.17)), which yields:

$$K_m = k u_* z. \tag{2.20}$$

Thus eddy transport is increasingly effective with height above the ground surface, and also increases linearly with u_*, and hence with $\bar{u}$.

2.4. Non-radiative Sensible Heat Exchanges in the Atmosphere near the Ground

Heat transfer by *convection* involves mixing of relatively large volumes of air at different temperatures. If in the atmosphere this mixing is due to forced or frictional turbulence (section 2.2), the resultant heat transfer is said to be by *forced convection*. When the mixing is caused by a difference in density that accompanies a temperature difference, the transfer is called *free* (or *natural*) *convection*. The latter mode of convective transfer dominates with strong lapse rates and small wind shear, such as would be found if dry ground were absorbing a high level of radiation under calm wind conditions. Forced convection on the other hand would predominate under small temperature gradients and strong wind shears. From eqn. (2.19), rate or shearing of the wind, which equals the velocity gradient, is proportional to u_*, and so to $\bar{u}$ very nearly. Thus cloudy skies and a high wind would ensure the predominance of forced convection. These conditions favour a near neutral atmosphere, so that we can take $K_h = K_m$, with K_m given by eqn. (2.20). Mean flux of sensible heat H could then be calculated using eqn. (2.14) under conditions defined by a measured mean temperature gradient $\partial \bar{T}/\partial z$ at height z, and friction velocity u_* (see Table 2.1).

But the question remains: What is a general criterion ensuring that convective transfer is fully forced or dominantly free; and over what range of atmospheric conditions are both modes of convection of comparable importance? An outline of answers to these questions will now be given.

We saw in discussing atmospheric stability (section 2.2) that negative vertical gradients of potential temperature $\partial \theta/\partial z$, or lapse conditions, tend to result in instability; and positive gradients, or inversion conditions, favour stability and the suppression of turbulence. (Near the ground we could replace $\partial \theta/\partial z$ by $\partial T/\partial z$ with no error significant for these purposes.) But whether or not turbulence will build up or subside is a dynamical problem, depending chiefly on the ratio of buoyancy forces (due to density differences) to frictional forces. Richardson[34] derived the criterion governing the growth or decay of turbulence in an atmosphere where temperature, and therefore density, varies with height. In his analysis he considered the degree of turbulence initially to be small, with flow bordering on the laminar state. For such conditions Richardson made the hypothesis that the kinetic energy of turbulence would increase or decrease according as the rate of energy supply from the eddy shear forces exceeded or was less than the rate at which work had to be done against gravity in raising masses of fluid—a characteristic feature of free convection. It is work extracted from the mean wind motion by eddy shearing stresses which maintains the energy of forced turbulence. This rate of extraction per unit volume is given by ($\tau \times$ rate of shear) or by ($\tau \times$ velocity gradient), which from eqn. (2.11) is proportional to $(d\bar{u}/dz)^2$.

Richardson concluded that a slightly turbulent motion would remain turbulent provided a non-dimensional ratio reflecting the relative rate of working of buoyancy and eddy shear forces, now known as the *Richardson number* R_i, was less than unity (i.e. fractional or negative). If R_i was greater than unity, turbulent flow would subside into a flow of laminar character. Alternative derivations of Richardson's criterion are given by Sutton[20, 21] which help bring out its physical significance. The Richardson number is given by:

$$R_i = \frac{g(\partial\theta/\partial z)}{\theta(\partial\bar{u}/\partial z)^2},\qquad(2.21)$$

where potential temperature θ may with small error be replaced by absolute air temperature T. The value of R_i depends on the height at which the gradients are specified (often taken as $1\cdot5$ m). It follows that in neutral conditions R_i is close to zero, when any turbulence present and convectional transfer of heat would be forced or frictional. The sign of R_i clearly depends on that of $(\partial\theta/\partial z)$, being positive under inversion conditions, when turbulence tends to be suppressed. Under lapse conditions R_i will be negative, and the greater the lapse rate in comparison with wind shear (i.e. the greater $-R_i$) the more likely it becomes that free convection will predominate over forced convection in heat transfer from the earth's surface. The existence on warm clear days of free convection currents up to considerable heights in the atmosphere, often referred to as "thermals", are used for soaring by birds and glider pilots alike.

Priestley[35] has shown that under temperature lapse conditions, convective transfer of heat in the lower atmosphere is effectively fully forced in the region $0 < (-R_i) < 0\cdot02$ approximately. Under conditions described by increasingly negative values of R_i, a rather sudden transition was found in the region of $R_i = -0\cdot02$ or $-0\cdot03$, beyond which free convection was dominant. Thus forced and free convection are of comparable importance only over a rather limited range of Richardson numbers, rather than over a wide range of conditions, as it was customary to believe. This information defines the range of conditions over which we may calculate convective heat transfer from the ground using the method appropriate for forced convection described at the beginning of this section.

For values of $(-R_i) > 0.02$ approximately, Priestley[36] has shown that the predominantly free convective transfer may be calculated from the temperature gradient at any particular height z with the equation:

$$H = \overset{*}{H} \rho_a c_p \left(\frac{g}{\theta}\right)^{\frac{1}{2}} \left|\frac{\partial \theta}{\partial z}\right|^{\frac{3}{2}} z^2 \qquad (2.22)$$

with the quantity $\overset{*}{H}$ a constant of value probably in the range 0·8 to 1·0.

For values of R_i between -0.02 and 0 (corresponding to forced convection), it may be proved that eqn. (2.22) can be regarded as an alternative to eqn. (2.14) (with eqns. (2.16) and (2.21)) for calculating turbulent heat loss into a near neutral atmosphere, provided $\overset{*}{H}$ is taken as $k^2 |R_i|^{-\frac{1}{2}}$.

In considering the thermal balance of isolated crop elements, such as a citrus fruit, convective heat transfer can be calculated using *surface* (or *film*) *heat transfer coefficients*. If the agricultural element can be approximated by a simple geometrical shape, appropriate values for such transfer coefficients can often be found from the engineering literature on heat transfer (Fishenden and Saunders[37]). Heat transfer across the boundary layer surrounding each surface is by a combination of conduction and convection if the boundary layer is turbulent, or conduction alone if it is laminar. The heat transfer coefficient h for a surface in a given situation is a measure of the net effective thermal conductance of the surface film, and is defined by:

$$H = h \Delta T,$$

where ΔT is the temperature difference between the surface and the bulk air outside the boundary layer.

Bibliography

1. SLATYER, R. O., and McILROY, I. C., *Practical Microclimatology*. Published by Commonwealth Scientific and Industrial Research Organization for UNESCO, 1961.

2. GEIGER, R., *The Climate Near the Ground*. Amplified 2nd ed. Harvard University Press, Massachusetts, 1959.
3. SMITH, W. O., and BYERS, H. G., The thermal conductivity of dry soils of certain of the great soil types, *Proc. Soil Sci. Soc. Amer.* **3**, 13 (1938).
4. WEST, E. S., The effect of soil mulch on soil temperature, *J. Coun. Sci. Industr. Res. Aust.* **5**, 236 (1933).
5. RIDER, N. E., A note on the physics of soil temperature, *Weather*, **12**, 241 (1957).
6. SHAW, B. T. (ed.), *Soil Physical Conditions and Plant Growth*. Academic Press, New York, 1952.
7. KEEN, B. A., *The Physical Properties of the Soil*. Longmans, London, 1931.
8. DE VRIES, D. A., The thermal conductivity of granular materials, *Bulletin de l'Institute International du Froid, Annexe*, **1952–1**, 115 (1952).
9. CALLENDER, H. L., and McLEOD, C. H., Observations of soil temperatures with electrical resistance thermometers, *Trans. Roy. Soc. Can.* 2nd. Series, **2**, 109 (1896–7).
10. SMITH, W. O., Thermal conductivities in moist soil, *Proc. Soil Sci. Soc. Amer.* **4**, 32 (1940).
11. GURR, C. G., MARSHALL, T. J., and HUTTON, J. T., Movement of water in soil due to a temperature gradient, *Soil Sci.* **74**, 335 (1952).
12. DE VRIES, D. A., Simultaneous transfer of heat and moisture in porous media, *Trans. Amer. Geophys. Un.* **39**, 909 (1958).
13. DE VRIES, D. A., A nonstationary method for determining thermal conductivity of soil *in situ*, *Soil Sci.* **73**, 83 (1952).
14. WEST, E. S., A study of the annual soil temperature, *Aust. J. Sci. Res.* Series A, **5**, 303 (1952).
15. CARSLAW, H. S., and JAEGER, J. C., *Conduction of Heat in Solids*. Oxford University Press, 1947.
16. DEACON, E. L., The measurement and recording of the heat flux into the soil, *Quart. J. R. Met. Soc.* **76**, 479 (1950).
17. PHILIP, J. R., The theory of heat flux meters, *J. Geophys. Res.* **66**, 571 (1961).
18. BROOKS, F. A., *An Introduction to Physical Microclimatology*. University of California, Davis, 1960.
19. SHAW, N., *Manual of Meteorology*. Vol. 2, 2nd ed. Cambridge University Press, London, 1936.
20. SUTTON, O. G., *Micrometeorology*. McGraw-Hill, New York, 1953.
21. SUTTON, O. G., *Atmospheric Turbulence*. 2nd ed. Methuen, London, 1954.
22. SUTTON, O. G., *The Science of Flight*. Penguin, Harmondsworth, 1949.
23. RICHARDSON, L. F., The supply of energy from and to atmospheric eddies, *Proc. Roy. Soc.* Series A, **97**, 354 (1920).
24. ECKERT, E. R. G., and DRAKE, R. M., *Heat and Mass Transfer*. 2nd ed. McGraw-Hill, New York, 1959.
25. TAYLOR, R. J., A linear, unidirectional anemometer of rapid response, *J. Sci. Instrum.* **35**, 47 (1958).
26. DYER, A. J., Measurements of evaporation and heat transfer in the lower atmosphere by an automatic eddy-correlation technique, *Quart. J. R. Met. Soc.* **87**, 401 (1961).

27. TAYLOR, R. J., and DYER, A. J., An instrument for measuring evaporation from natural surfaces, *Nature, Lond.* **181,** 408 (1958).
28. PASQUILL, F., The aerodynamic drag of grassland, *Proc. Roy. Soc.* Series A, **202,** 143 (1950).
29. PASQUILL, F., Eddy diffusion of water vapour and heat near the ground, *Proc. Roy. Soc.* Series A, **198,** 116 (1949).
30. SWINBANK, W. C., The measurement of vertical transfer of heat and water vapour by eddies in the lower atmosphere, *J. Met.* **8,** 135 (1951).
31. RIDER, N. E., Eddy diffusion of momentum, water vapour, and heat near the ground, *Phil. Trans.* Series A, **246,** 481 (1954).
32. PENMAN, H. L., and LONG, I. F., Weather in wheat: an essay in micro-meteorology, *Quart. J. R. Met. Soc.* **86,** 16 (1960).
33. DEACON, E. L., Vertical diffusion in the lowest layers of the atmosphere, *Quart. J. R. Met. Soc.* **75,** 89 (1949).
34. RICHARDSON, L. F., The supply of energy from and to atmospheric eddies, *Proc. Roy. Soc.* Series A, **97,** 354 (1920).
35. PRIESTLEY, C. H. B., Free and forced convection in the atmosphere near the ground, *Quart. J. R. Met. Soc.* **81,** 139 (1955).
36. PRIESTLEY, C. H. G., *Turbulent Transfer in the Lower Atmosphere*. University Press, Chicago, 1959.
37. FISHENDEN, M., and SAUNDERS, O. A., *An Introduction to Heat Transfer*. Oxford University Press, London, 1957.

CHAPTER 3

The Physical Environment of Agriculture: Part III

THIS chapter is concerned with water in the atmosphere and with the theoretical background to some methods of measuring evaporation from the earth's surface. This subject will be taken further, and its more practical aspects considered, in section 7.1.

3.1. Humidity

WATER VAPOUR

From a meteorological viewpoint, the atmosphere is a mixture of (a) *dry air* (of almost constant composition but with density decreasing with height), and (b) *water vapour*. It is vital for photosynthesis that a component of dry air is carbon dioxide, and we shall later be considering the important role of water vapour in the plant microclimate; but water vapour and carbon dioxide together only amount to 1 or 2 per cent as components of the atmosphere, with carbon dioxide only about 0·03 per cent by volume on average.

As described by Dalton's law of partial pressures, the pressure exerted by water vapour is independent of the pressure exerted by other atmospheric gases. Furthermore, provided conditions are such that there is no condensation or evaporation taking place in the volume under consideration, the behaviour of water vapour is similar to any other gas, which to a close approximation can be represented by the equation of state of an ideal gas. The vapour pressure of water (e, dyne cm^{-2}) under these conditions can thus be represented by:

$$e = \frac{\rho_v}{M_w} R_u T, \qquad (3.1)$$

where ρ_v is the density of water vapour (g cm^{-3}) and M_w ($= 18$ g mole^{-1}) its molecular weight. R_u is the universal gas constant ($8 \cdot 31 \times 10^7$ erg mole^{-1} deg^{-1}K). T, the absolute temperature of the water vapour, may be taken as the temperature of the air of which the vapour is one component. The density of water vapour, ρ_v, is referred to as the *absolute humidity* of the atmosphere. (As ρ_v in g cm^{-3} is a small quantity, it is commonly expressed in units of g m^{-3}.)

SATURATION

Consider an enclosed container at constant temperature which is partially filled with pure water bounded by a plane liquid surface. Because of the considerations behind Dalton's law, the equilibrium established is the same whether or not there is any other gas in the space above the liquid. At equilibrium the number of water molecules leaving the liquid surface per second due to thermal excitation will be the same as the number of vapour molecules being recaptured per second on striking the liquid surface. When this equilibrium is attained the vapour is said to be *saturated*. The vapour pressure is then the maximum possible for that particular temperature under these conditions, and is known as the *saturation vapour pressure* (s.v.p.). (If the vapour is in equilibrium with a convex liquid surface, and in other specially created circumstances, this value can be exceeded; but under normal circumstances the s.v.p. is a unique function of temperature.) The rapid increase of the s.v.p. of water with temperature is illustrated by the curve in Fig. 22. (Tabulated values are given by Kaye and Laby[1] and the Smithsonian Institution,[2] for example.) In this figure vapour pressure is expressed in the two pressure scales in common use in humidity measurements. The scale based on the pressure exerted by a height of mercury is not absolute (since it depends on the value adopted for g), and it does not belong to the c.g.s. system of units, where the absolute unit of pressure is 1 dyne cm^{-2}. Atmospheric pressure at sea level is of the order of 10^6 dyne cm^{-2} (the "standard" atmospheric pressure adopted being that exerted by 760 mm of mercury with $g = 980 \cdot 7$ cm sec^{-2} corresponding to $1 \cdot 013 \times 10^6$ dyne cm^{-2}). In meteorology a multiple of the absolute unit known as a millibar (mb) is used, where the following relations hold:

$$1 \text{ bar} = 10^6 \text{ dyne cm}^{-2}$$
$$1 \text{ millibar (mb)} = 10^3 \text{ dyne cm}^{-2}$$
$$= 0 \cdot 750 \text{ mm of mercury.}$$

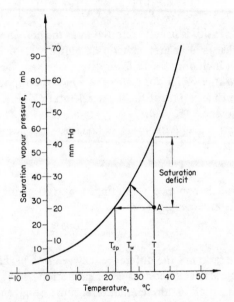

FIG. 22. The saturation vapour pressure of water shown as a function of temperature (after Penman[3]).

At temperatures below 0°C, the s.v.p. is somewhat different if the vapour is in equilibrium with ice rather than water—the case illustrated in Fig. 22. Since at the boiling point the s.v.p. is equal to the external pressure, the curve in Fig. 22 could alternatively be interpreted as the variation of boiling point (abscissa) with external atmospheric pressure (ordinate).

METHODS OF EXPRESSING PARTIAL SATURATION

The vapour pressure e and vapour density ρ_v already introduced are two ways of measuring atmospheric humidity. A useful form of eqn. (3.1) with e in *millibars* and T in °K is:

$$\left.\begin{aligned} \rho_v &= 217\frac{e}{T} \quad (\mathrm{g\,m^{-3}}) \\[4pt] &= 217 \times 10^{-6}\frac{e}{T} \quad (\mathrm{g\,cm^{-3}}). \end{aligned}\right\} \tag{3.2}$$

Vapour density is usually referred to as the *absolute humidity*.

Other measures at atmospheric humidity are sometimes more convenient, including the following:

The *saturation deficit* for a space in which the vapour pressure e is less than the s.v.p. e_s at the air temperature T (Fig. 22) is given by the differences $(e_s - e)$.

The *relative humidity* h_r is defined as the non-dimensional ratio ($\leqslant 1$);

$$h_r = \frac{e}{e_s}. \tag{3.3}$$

This fraction is commonly multiplied by 100 giving the relative humidity as a percentage:

$$\text{r.h.} = 100h_r = \frac{100e}{e_s}.$$

Neither the ratio nor the difference between e and e_s allows e or e_s to be calculated. Thus besides h_r, or saturation deficit, some other information, such as air temperature, must be given before any humidity condition is fully defined.

The *dew-point temperature* T_{dp} can also be illustrated by reference to Fig. 22. Let a parcel of air whose humidity condition is represented by A be cooled at constant pressure without any gain or loss of water vapour. Radiative cooling of air on a calm clear night can fulfil these conditions provided there is no dew formation. The horizontal line to the left of A represents the state path of this change. This state path eventually intersects the s.v.p. curve at such a temperature that the vapour is saturated. This temperature T_{dp} is called the dew point since condensation of dew would form on any surface in contact with the vapour if it were cooled below it. In other words the dew point corresponding to any degree of saturation is that temperature to which the vapour must be reduced in an *isobaric* process to become saturated.

Wet-bulb temperature T_w. Consider the cooling of air produced when it flows over wet surfaces. If point A of Fig. 22 represents the initial state of such an air flow, the evaporation of moisture into the air stream will both increase the vapour pressure and reduce the temperature of the air stream by extracting from it the latent heat required for evaporation. Such a change can be represented by the state path inclined upwards and to the left of A, which intersects the s.v.p. curve at a temperature which will depend somewhat on the conditions of the air flow. This temperature decreases as flow rate is increased, but it is found in practice that provided the ventilation rate is greater than about $2\,\mathrm{m\,sec^{-1}}$, a small wet surface achieves a steady lower temperature which is rather insensitive to further increases in flow rate provided this small wet surface is to some extent thermally isolated from its surroundings. This steady lower temperature is defined as the wet-bulb temperature T_w corresponding to state A.

The *mixing ratio x* of moist air is the ratio of the mass of water vapour to the mass of *dry air* with which the water vapour is associated (i.e. the mass of water vapour per unit *mass* of dry air). The *specific humidity q*, defined as the mass of water vapour per unit mass of *moist* air, is very little different in magnitude from the mixing ratio for atmospheric humidities. From the ideal gas equation (3.1) it follows that the density of a gas is proportional to:

$$\frac{(\text{pressure})(\text{molecular weight})}{\text{absolute temperature}}.$$

Let p_a be the total atmospheric pressure, and M_a the molecular weight of dry air.

Then
$$x = \rho_v/\rho_a', \quad \text{where } \rho_a' = \text{dry air density}$$
$$= \frac{eM_w}{(p_a - e)M_a}.$$

A mean value of M_a, the molecular weight of air, is close to 28.9, the value adopted for (M_w/M_a) usually being 0.622.

Thus

$$x = 0.622\frac{e}{p_a - e}$$

$$= 0.622\frac{e}{p_a} \quad \text{very closely,} \tag{3.4}$$

provided e is small compared with p_a, as it is in natural climates. It will be noticed that the mixing ratio is independent of temperature. It is therefore constant for any volume of air provided there is no condensation or evaporation, and until it mixes with air of different mixing ratio. This measure of humidity is of particular use in energy balance studies, and will be used in deriving the equation of the wet- and dry-bulb hygrometer.

WET- AND DRY-BULB HYGROMETER

The most commonly used type of instrument in accurate agricultural humidity investigations is at present the wet- and dry-bulb type of hygrometer, usually referred to as a psychrometer. Penman's[3] monograph of humidity, on which this section is much dependent, describes other methods of humidity measurement. Because of the varied and extensive need for convenient hygrometers, instrumental development is active (see Wylie[4] and Cutting et al.,[5] for example).

In the wet- and dry-bulb hygrometer, the bulb refers to the mercury bulb in mercury-in-glass type thermometers. In the instrument one bulb is kept wet, perhaps by covering it with a close fitting muslin sheath, kept moist by a wick connecting it to a reservoir of distilled water. Water will evaporate from the wet bulb (unless the air is saturated) so that it would register a lower temperature than an adjacent thermometer with a dry bulb. Clearly, mercury-in-glass thermometers can be replaced by any other types of thermometer, such as electrical resistance or thermocouple types (McIlroy[6] and Powell[7] give examples).

In developing a theory relating vapour pressure to temperature depression of the wet bulb, various assumptions may be made. It is assumed that a steady state exists, and that all the heat required to vaporize water from the wet bulb is taken from air flowing over the wet bulb. If the wet bulb is at a lower temperature than its surroundings it can gain heat by radiation exchange with them, and by conduction along supports. Such sources of error have been considered by Monteith.[8] Since by suitable design these errors can usually be kept small, they will be neglected in the theory. The design problems in keeping such errors small are most severe with day-time humidity measurements in the field, due to the intense radiation, and some sort of compromise between adequate shielding from radiation, representative exposure and suitable ventilation has to be made in these conditions.

Following Spilhaus,[9] let the heat given up per unit time by the incident air in cooling the wet bulb be associated with a mass m_1 of dry air which is cooled from T to the wet-bulb temperature T_w. This amount of heat is thus given by:

$$m_1 c_p (T - T_w),$$

where c_p is the specific heat of dry air at constant pressure. The correction to c_p due to the incident air not being dry is small for atmospheric conditions, and has been considered by Whipple.[10]

Now consider evaporation from the wet bulb. Let the mass of water evaporated per unit time be associated with a mass m_2 of dry air. Let the water content of the incident air be defined by a mixing ratio x (grams per gram of dry air); then the air displaced from around the wet bulb will have a mixing ratio x_w, the saturation value of the mixing ratio at temperature T_w. The mass of water evaporated per second from the wet bulb is thus $m_2 (x_w - x)$ grams. On the assumption that all the heat required to vaporize this mass of water comes from the air, we equate:

$$m_2 L (x_w - x) = m_1 c_p (T - T_w)$$

or

$$x = x_w - \frac{m_1 c_p}{m_2 L} (T - T_w), \qquad (3.5)$$

where L is the latent heat of vaporization of water in the neighbourhood of temperature T_w, assumed constant.

At least for relatively low wind speeds (just how low depending on thermometer size), the mass ratio m_1/m_2 is dependent on this speed, though it is approximately constant for higher wind speeds. The common classical assumption (associated with the name of August) is to take $m_1 = m_2$ (Whipple[10]). With this assumption, and using eqn. (3.4), eqn. (3.5) can be recast in terms of vapour pressures thus:

$$\frac{0{\cdot}622e}{p_a - e} = \frac{0{\cdot}622e_w}{p_a - e_w} - \frac{c_p}{L}(T - T_w).$$

With the approximation (which is a good one in agricultural situations) that

$$p_a - e = p_a - e_w = p_a,$$

it follows that

$$\begin{aligned} e &= e_w - p_a \frac{c_p}{0{\cdot}622L}(T - T_w) \\ &= e_w - p_a A(T - T_w), \end{aligned} \tag{3.6}$$

where A is called the "psychrometric constant" (Smithsonian Institution[2]), though this term is sometimes used of $(p_a A)$ (Middleton and Spilhaus[11]). A is dependent on wind speeds when these are low, the International Meteorological Organization in 1947 recommending the use of air speeds between 4 and 10 m sec^{-1} for mercury-in-glass type thermometers (Smithsonian Institution, loc. cit.). In agricultural microclimatic investigations the provision of such an air blast may appreciably alter the environment it is desired to measure. Consequently, psychrometers in such investigations are often unaspirated. A separate measurement of wind speed in a similar exposure enables appropriate values of "psychrometric constant" to be used.

Putting $p_a = 1000$ mb, $L = 586$ cal g^{-1} (at 20°C), $c_p = 0{\cdot}24$ cal g^{-1} deg^{-1}C in eqn. (3.6) gives:

$$e = e_w - 0{\cdot}66(T - T_w) \quad (e \text{ in mb}, T \text{ in } °C).$$

Marvin[12] provides a most convenient form of general psychrometric tables applicable to full ventilation. The tables give dew-point temperature for measured values of air temperature and wet-bulb depression, and are presented for a range of atmospheric pressures p. Routine humidity observations are often made in a raised louvred box known as a "Stevenson screen" (Middleton and Spilhaus[11]), in which the only hygrometer ventilation is due to natural air movement through the screen. The appropriate psychrometric constant is higher in this case than with full ventilation, and is also less certain due to the variability of natural wind. (In consulting psychrometric tables it is obviously important to ensure that they apply to the ventilation employed in measurement.)

Two advantages accompanying the use of small thermometer elements in psychrometry are of particular importance in agricultural field investigations. The first is that the degree of shielding necessary to prevent appreciable error due to absorption of short-wave radiation decreases with thermometer size. This is because radiation exchange is proportional to surface area, but convective heat loss per unit area increases with decrease in size. For example, for cylindrical bodies of diameter d, heat loss *per unit area* for a given temperature difference between the body and air is found to be approximately proportional to $d^{-0.4}$ under aspirated conditions with forced convection, and to be proportional to $d^{-0.25}$ with free or natural convection (Brown and Marco[13]). For small temperature differences radiation transfer is closely proportional to temperature difference. The surface area of cylindrically shaped objects, and hence radiation exchange, is proportional to d. Thus, for a given temperature difference, and for unit area of cylindrically shaped objects (which is a good approximation to most practical wet- and dry-bulb elements), we have the ratio:

$$\frac{\text{convective heat transfer}}{\text{radiant heat transfer}} \propto \frac{d^{-0.4}}{d} \propto \frac{1}{d^{1.4}}.$$

Thus, for sufficiently small diameters, the heat balance of any object will be dominated by convective heat transfer, radiation exchange being of decreasing significance with decreasing size.

Similar considerations are also of importance in understanding the heat balance of plants and animals (Monteith[14]).

The second advantage accruing from the use of small thermometer elements is that the need for ventilation is very much less. This also depends on the fact illustrated above for cylindrical shaped bodies, for which natural convective heat transfer per unit area increases proportionally to $d^{-0.25}$. The need for additional heat transfer by forced flow thus decreases with size, natural convection alone eventually becoming adequate to transfer the heat required to vaporize water from the wet bulb at the equilibrium rate corresponding to "full" ventilation. As mentioned above, adequate ventilation of large thermometer elements is an undesirable disturbance in microclimatic research. Rider[15] and McIlroy[16] have described designs of field microclimate psychrometers for which such objections are reduced to a minimum.

To illustrate this effect of size on ventilation requirement, Powell[7] investigated the behaviour of a psychrometer employing thermojunctions of 44 s.w.g. (0·081 mm diameter) nichrome and constantan wires joined end to end. A wet cotton wrapping extending about 0·25 in. either side of a junction provided the "wet bulb". Results obtained for relative humidity with the elements exposed in quite still air were only about 2 or 3 per cent greater than with full ventilation, when psychrometric tables appropriate to full ventilation were used in both instances.

A secondary standard instrument often used in checking the performance of psychrometers is the Assman hygrometer (Middleton and Spilhaus[11]), in which air is drawn past shielded mercury-in-glass thermometer elements by means of a fan.

3.2. Evaporation

Water can be evaporated into the air from plant, soil and water surfaces. Evaporation from plant surfaces of water which has been transmitted from the soil through the plant is often referred to as *transpiration*, but the phase change implied by the term evaporation is physically identical wherever it takes place.

For evaporation from a surface to continue there are three

physical requirements. Firstly, there must be a supply of heat to provide the quite large latent heat of vaporization ($590 \, \text{cal g}^{-1}$ around $15°C$). Secondly, the vapour pressure in the overlying air must be maintained at less than that at the evaporating surface, since evaporation (a *net* transfer of water vapour) is zero if there is no gradient in vapour pressure. The third requirement is simply that water must continue to be available for evaporation, this being a limiting factor under dry conditions. Evaporation, through its latent heat, is thus a component in the energy balance. It is also part of a transport process whereby rain water and precipitation generally is returned to the atmosphere. The evaporation process is so linked in nature that it can be measured using either an "energy balance" approach, based on the first consideration, or on a "sink-strength" or "aerodynamic" treatment, based on the second. Yet a third approach is to measure all components in the water-balance equation (1.11) except E.

A convenient critical summary of the methods available for measuring natural evaporation has been given by McIlroy.[16] Greater practical and theoretical detail is given by Slatyer and McIlroy.[17] Excellent reviews of the subject of evaporation are available, for example by Penman[18, 19] and by Deacon, Priestley and Swinbank.[20] Such sources and their references give more information, and the following discussion will be limited to an understanding of physical principles behind the aerodynamic and energy balance methods, and a combined method approach. The physical basis of the direct eddy correlation method, and of the water-balance method has already been discussed in outline (sections 2.3 and 1.3 respectively).

AERODYNAMIC TURBULENT–TRANSFER APPROACH

The validity of this method depends on three propositions for which there is experimental support under fully turbulent near neutral atmospheric conditions (Rider[21]). These are:

(a) that the mean wind speed varies logarithmically with height as described by eqn. (2.18);

(b) that eddy shear stress is constant with height (eqn. (2.13)); and

(c) that the eddy diffusivity for water vapour K_w is equal to that for momentum K_m (eqn. (2.16)).

The conditions and restrictions for these propositions to be supported experimentally were outlined in section 2.3, the most important being a near-neutral atmosphere and an open uniform site.

The equation used in evaporation measurement may be derived as follows (Thornthwaite and Holzman[22]):

Substituting for friction velocity u_* from eqn. (2.19) into eqn. (2.20) we get:

$$K_m = k^2 z^2 \frac{d\bar{u}}{dz}. \tag{3.7}$$

Assuming $K_h = K_m$, with K_m given by eqn. (3.7), it follows from eqn. (2.15) that water vapour flux E is given by:

$$E = -k^2 z^2 \frac{d\bar{u}}{dz} \cdot \frac{d\bar{\rho}_v}{dz}.$$

Carrying out a double integration between two observational heights z_1 and z_2 yields an equation in which surface properties are eliminated. Thus:

$$\int_{z_1}^{z_2} \int_{z_1}^{z_2} E \left(\frac{dz}{z} \right)^2 = k^2 \int_{z_1}^{z_2} d\bar{u} \int_{z_1}^{z_2} d\bar{\rho}_v.$$

Provided E is independent of z, i.e. assuming no horizontal advection (section 1.4) of water vapour, it follows that:

$$E = \frac{k^2 (\bar{u}_2 - \bar{u}_1)(\bar{\rho}_{v1} - \bar{\rho}_{v2})}{(\ln z_2/z_1)^2}. \tag{3.8}$$

Following Pasquill[23] this may be simplified, using eqn. (3.1) to:

$$E = B\bar{u}_2(\bar{e}_1 - \bar{e}_2),\qquad(3.9)$$

where

$$B = \frac{k^2 M_w(1 - \bar{u}_1/\bar{u}_2)}{R_u T(\ln z_2/z_1)^2}\qquad(3.10)$$

is very nearly a constant for any pair of observational heights above a surface of particular roughness in neutral conditions, as follows from eqn. (2.18). Equation (3.9) thus shows vapour flux to be proportional to mean wind speed, and to the vapour pressure difference between any two measurement heights z_1 and z_2. This difference is a measure of the gradient in vapour pressure necessary for vapour transfer, and mean wind speed is a measure of the efficiency of turbulent mixing. The importance of these two factors in controlling evaporation was realized over a century ago by Dalton. Evaporation equations, developed either empirically or using boundary layer data (Penman[18]) but based on these ideas were in use prior to the aerodynamic approach outlined above. Such equations have usually been expressed in terms of the difference between surface and bulk air values of vapour pressure, in the form:

$$LE = f(\bar{u})(e_o - e),\qquad(3.11)$$

where $f(\bar{u})$ is an empirically determined function of mean wind speed, often found to be well represented by:

$$f(\bar{u}) = a(1 + b\bar{u})\qquad(3.12)$$

with a and b constants.

Equations of this type are useful in predicting evaporation from water surfaces, where surface vapour pressure e_o may be taken as the saturation vapour pressure at surface temperature, itself a quantity which can be measured fairly readily. However, this is not true for even partially dry surfaces, even assuming that an effective average surface temperature could be measured. Because of the very large gradients in the boundary layer sheathing, ground or plant surfaces of all properties including vapour pressure, e_o is an extremely difficult quantity to measure directly. Thus such equations have been developed chiefly with evaporation from water.

With taller uniform vegetation, it is necessary to reduce z (measured from the ground surface) in eqn. (3.10) by the experimentally determined zero-plane displacement mentioned in section 2.3. For this purpose it is preferable to measure $\bar{u}$ for at least three heights z, when d can be found by trial as that height which gives the best linear plot of $\bar{u}$ against $\ln(z - d)$. It will have to be redetermined as the crop grows, and will vary with wind speed in pliable crops (see Rider[24] for measurements on oats).

Whilst the use of eqns. (3.9) and (3.10) to calculate evaporation rate from soil or crops gets over the problem of determining e_o necessary in the Dalton approach illustrated by eqn. (3.11), the restriction to near-neutral conditions is a limitation which so far has been overcome only in a semi-empirical manner. High accuracy is necessary in the measurement of $\bar{u}$, and especially in the difference $(e_1 - e_2)$ because of the small magnitude of gradients in the well-mixed turbulent air above the surface boundary layer. Furthermore, long-term mean values for these quantities cannot be used without introducing serious errors into E, so computations at intervals of the order of an hour are necessary. For such reasons the method is likely to be used chiefly in short period research investigations unless automatic data logging is employed together with machine computation.

ENERGY BALANCE METHOD

All components in the energy balance equation at the ground surface (eqn. (1.8)) can be measured directly, as described in section 1.2 for R_N, and in section 2.1 for G. In employing the energy balance approach, the net energy available at the surface, $R_N - G$, has usually been partitioned between heating the air H and evaporation LE by considering their ratio, an approach introduced by Bowen.[25] Working on evaporation from open water, Bowen expressed the ratio H/LE in terms of surface and screen height properties. However, from the transport equations (2.14) and (2.15), it follows that the so-called Bowen ratio (i.e. H/LE) can be expressed as:

$$\beta = \frac{H}{LE} = \frac{c_p K_h (\partial \bar{T}/\partial z)}{L K_w (\partial \bar{q}/\partial z)}$$

$$= \gamma \frac{\Delta \bar{T}}{\Delta \bar{q}}, \tag{3.13}$$

assuming $K_h = K_w$, and writing $\gamma = c_p/L$, a ratio appearing in the wet- and dry-bulb psychrometer equation (3.6), whose magnitude varies slightly with temperature, being about $4 \cdot 2 \times 10^{-4}$ deg^{-1}C at ordinary temperatures. In eqn. (3.13) $\Delta \bar{T}$ and $\Delta \bar{q}$ are differences over the same height interval of mean air temperature and specific humidity respectively. Thus β can be measured experimentally. With a moist surface β is usually between zero and $0 \cdot 2$. If there is advection of sensible heat H can be negative (i.e. downward), and so $\beta < 0$.

From eqn. (1.10):

$$\frac{R_N - G}{LE} = 1 + \beta.$$

Thus

$$E = \frac{R_N - G}{L(1 + \beta)}, \tag{3.14}$$

which, together with eqn. (3.13) for β, can be used to determine E. Especially if β is not large (so that the accuracy demanded in measurements of $\Delta \bar{T}$ and $\Delta \bar{q}$ is not stringent), and provided these measurements are made at suitably low levels for the assumption $K_h = K_w$ to be reasonable, this approach is capable of good accuracy. As with the aerodynamic method an analysis about each hour is necessary for a daily value of E to be reliable.

COMBINATION METHODS OF EVAPORATION ESTIMATION

As mentioned above, the difficulty in using a Dalton type of equation such as eqn. (3.11) is that vapour pressure at the surface

e_o is difficult to measure. Where a known relation between e_o and surface temperature T_o exists, the energy balance equation may be combined with a Dalton type equation to eliminate these two variables. This is the case for water surfaces or land surfaces sufficiently close to saturation to assume e_o is the saturation vapour pressure at temperature T_o. For such conditions Penman[26] and Ferguson[27] independently proposed similar methods. Penman's approach is outlined:

Equation (3.11) is

$$LE = (e_o - e)f(u),$$

where e is the vapour pressure "in the air". Because of the small gradients outside the surface boundary layer the exact height at which e is specified is unimportant, the majority of the vapour pressure drop taking place across the surface boundary layer.

A quantity E_a is now introduced, defined by replacing e_o in the expression for E by e_a, the symbol used by Penman (loc. cit.) for the saturation vapour pressure at air temperature, denoted previously by e_s (p. 72). Then:

$$LE_a = (e_a - e)f(u).$$

Thus

$$\frac{E_a}{E} = 1 - \left[\frac{e_o - e_a}{e_o - e} \right] = 1 - \varphi \qquad \text{say.} \qquad (3.15)$$

As mentioned above, this approach assumes evaporation to be either from water or a saturated surface, as will become explicit below. For such conditions Penman assumed the ground heat flux term G to be zero, or more strictly that its average effect over periods greater than several days was negligible. Thus from the energy balance equation (3.14), putting $G = 0$:

$$\frac{R_N}{LE} = 1 + \beta.$$

As originally introduced by Bowen, the Bowen Ratio is:

$$\beta = \gamma' \frac{T_o - T_a}{e_o - e},$$

where γ' is different to the γ in eqn. (3.13), and T_a is air temperature.

Thus

$$\frac{R_N}{LE} = 1 + \gamma'\left(\frac{e_0 - e_a}{e_0 - e}\right)\bigg/\left(\frac{e_0 - e_a}{T_0 - T_a}\right) \qquad (3.16)$$

Now the following approximation is introduced for the last term on the right-hand side of eqn. (3.16):

$$\frac{e_0 - e_a}{T_0 - T_a} = \left(\frac{de}{dT}\right)_{T = T_a} = \Delta \qquad \text{say.}$$

For this to be even approximately true it is clear that the vapour pressure at the surface must be the saturation value. From this and from eqn. (3.15), eqn. (3.16) can be written:

$$\frac{R_N}{LE} = 1 + \frac{\gamma'}{\Delta}\varphi.$$

Substitution of $\varphi = 1 - E_a/E$ in this, with algebraic rearrangement leads to:

$$LE = \frac{(\Delta/\gamma')R_N + LE_a}{(\Delta/\gamma') + 1} \quad (\text{cal cm}^{-2}\,\text{sec}^{-1}), \qquad (3.17)$$

which can be regarded as representing a weighting of net radiation supply and aerodynamic effects in evaporation. The ratio Δ/γ' is a dimensionless temperature-dependent quantity.

R_N should be measured directly with a net radiometer, but if this is not possible empirical formulae like eqns. (1.6) and (1.7) may be used, accepting a greater possible inaccuracy. For an open water surface Penman has used the following expression for E_a, with LE_a in units of millimetres of water evaporated per day (see Appendix for conversion factors):

$$LE_a = 0{\cdot}35(e_a - e)(0{\cdot}5 + u_2/100) \quad (\text{mm day}^{-1}), \qquad (3.18)$$

where e_a = saturation vapour pressure at mean air temperature (mm of mercury),

   e = mean vapour pressure in air, or the saturation vapour pressure at the dew point (mm of mercury), and

   u_2 = mean horizontal wind velocity in miles day^{-1}, measured at 2 m above ground level.

Penman's approach in calculating evaporation from vegetation is firstly to calculate the evaporation from an open-water surface theoretically substituted for the surface of interest. Denote this by E_w. (If empirical equations like eqns. (1.6), (1.7) and (3.18) are used for R_N and E_a, only standard meteorological data are required to calculate E_w.) The ratio of transpiration rate E_T to E_w is then determined experimentally for the situation, season, and crop to which the results are to be employed. If the vegetation always has plenty of water available, Penman found the ratio $f = E_T/E_w$ to have a simple seasonal fluctuation in the British Isles. This approach has been found to give agreement with hydrologic balance data for catchment areas (Penman,[28] E.A.A.F.R.O.[29]). If soil moisture deficit is sufficient to restrict transpiration the use of simple seasonal ratios f is not applicable. Assuming roots can exploit a soil depth yielding an estimable amount of water, Penman[30] and Slatyer[31] have considered ways of taking such restrictions on evaporation into account.

Bibliography

1. KAYE, G. W. C., and LABY, T. H., *Tables of Physical and Chemical Constants*. 11th ed. Longmans, London, 1956.
2. *Smithsonian Meteorological Tables*, 6th rev. ed., The Smithsonian Institution, Washington, 1951.
3. PENMAN, H. L., *Humidity*. Chapman & Hall on behalf of Institute of Physics, 1955.
4. WYLIE, R. G., A new absolute hygrometer of high accuracy, *Nature, Lond.* **175**, 118 (1955).
5. CUTTING, C. T., JASON, A. C., and WOOD, J. L., A capacitance–resistance hygrometer, *J. Sci. Instrum.* **32**, 425 (1955).
6. McILROY, I. C., A sensitive temperature and humidity probe, *Aust. J. Agric. Res.* **6**, 196 (1955).
7. POWELL, R. W., The use of thermocouples for psychrometric purposes, *Proc. Phys. Soc. Lond.* **48**, 406 (1936).
8. MONTEITH, J. L., Error and accuracy in thermocouple psychrometry, *Proc. Phys. Soc. Lond.* Series B, **67**, 217 (1954).
9. SPILHAUS, A. F., A study of the aspiration psychrometer, *Trans. Roy. Soc. S. Afr.* **24**, 185 (1937).
10. WHIPPLE, F. J. W., The wet-and-dry-bulb hygrometer, The relation to theory of the experimental researches of Awberry and Griffiths, *Proc. Phys. Soc. Lond.* **45**, 307 (1933).

11. MIDDLETON, W. E. K., and SPILHAUS, A. F., *Meteorological Instruments*. 3rd ed., revised. University of Toronto Press, 1953.
12. MARVIN, C. F., *Psychrometric Tables*. U.S. Dept. of Commerce, Weather Bureau, No. 235, 1941.
13. BROWN, A. I., and MARCO, S. M., *Introduction to Heat Transfer*. 2nd ed. McGraw-Hill, New York, 1951.
14. MONTEITH, J. L., Micro-meteorology in relation to plant and animal life, *Proc. Linn. Soc. Lond.* **171**, 71 (1960).
15. RIDER, N. E., Water loss from various land surfaces, *Quart. J. R. Met. Soc.* **83**, 181 (1957).
16. MCILROY, I. C., The measurement of natural evaporation, *J. Aust. Ins. Agric. Sci.* **23**, 4 (1957).
17. SLATYER, R. O., and MCILROY, I. C., *Practical Microclimatology*. Published by Commonwealth Scientific and Industrial Research Organization for UNESCO, 1961.
18. PENMAN, H. L., Evaporation in nature, *Rep. Prog. in Phys.* **11**, London, Physical Society, 366 (1948).
19. PENMAN, H. L., Evaporation: an introductory survey. *Neth. J. Agric. Sci.* **4**, 9 (1956).
20. DEACON, E. L., PRIESTLEY, C. H. B., and SWINBANK, W. C., Evaporation and the water balance, *Climatology Reviews of Research* (Arid Zone Research, No. 10) *UNESCO* (1958).
21. RIDER, N. E., An account of the development of the aerodynamic method for the evaluation of natural evaporation conducted in Great Britain from 1947 to 1953, *L'Association Internationale d'Hydrologie*, Publication No. 38 (1954).
22. THORNTHWAITE, C. W., and HOLZMAN, B., The determination of evaporation from land and water surfaces, *Mon. Weath. Rev., Wash.* **67**, 4 (1939).
23. PASQUILL, F., Some estimates of the amount and diurnal variation of evaporation from a clayland pasture in fair spring weather, *Quart. J. R. Met. Soc.* **75**, 249 (1949).
24. RIDER, N. E., Evaporation from an oat field, *Quart. J. R. Met. Soc.* **80**, 198 (1954).
25. BOWEN, I. S., The ratio of heat losses by conduction and by evaporation from any water surface, *Phys. Rev.* **27**, 779 (1926).
26. PENMAN, H. L., Natural evaporation from open water, bare soil and grass, *Proc. Roy. Soc.* Series A, **193**, 120 (1948).
27. FERGUSON, J., The rate of natural evaporation from shallow ponds, *Aust. J. Sci. Res.* Series A, **5**, 315 (1952).
28. PENMAN, H. L., The water balance of the Stour catchment area, *J. Instn. Wat. Engrs.* **4**, 457 (1950).
29. E.A.A.F.R.O., Hydrological effects of changes in land use in some East African catchment areas, *E. Afr. Agric. For. J.* **27**, (entire number) (1962).
30. PENMAN, H. L., The dependence of transpiration on weather and soil conditions, *J. Soil Sci.* **1**, 74 (1950).
31. SLATYER, R. O., Agricultural climatology of the Katherine Area, N.T., C.S.I.R.O. *Division of Land Research and Regional Survey*. Technical Paper No. **13**, (1960).

CHAPTER 4

Some Physical Aspects of Soils

THE important question of how soil is formed, initially from the disintegration of rock by natural chemical and physical means known as *weathering*, but also involving the activities of diverse biological populations, is introduced by Comber,[1] and discussed in more detail by Robinson[2] and Russell[3] among others. Because this topic of soil formation and development is fully discussed elsewhere this important field of enquiry will receive no further attention here, except to comment on the meaning of the term "soil" in agriculture, and to mention the origin of the fine clay fraction of soil.

At least as used in an agricultural context, the term "soil" includes more than the complex of rock minerals, however far the weathering process may have gone. The profound influences of biological populations such as soil microorganisms and animals are recognized in this use of the word. Green plants furnish the organic matter which is the food supply of these populations. Without this decomposition of organic matter, which, since it ultimately yields carbon dioxide and water is effectively the reverse of photosynthesis, all the carbon dioxide in the atmosphere would be exhausted in a few decades. Microorganisms are now recognized as intimately associated with the production of the dark coloured and very finely divided organic material referred to as *humus*, which is a general component of surface soil distinct from mineral material.

The very finely divided mineral material in soil is referred to as clay. Clay is an important example of secondary minerals, being formed by very slow weathering processes, estimates of annual formation being of the order of 10^{-6} g per g of parent material.

Clay, together with the primary rock material and the uncombined oxides and carbonates (such as Fe_2O_3, SiO_2, $CaCO_3$) form the three main types of inorganic material in soil. The properties and proportion of this finely divided clay fraction have a very important role in both the physical and chemical properties of soil, and this soil fraction is the topic of section 4.2.

Both the nature of the parent material upon which soil develops, and the climate and topography through their effects for example on the downward movement of soluble material (or *leaching*), interact with other factors to produce a great variety of soil types whose properties often vary considerably with depth from the soil surface. Clarke[4] and the Soil Survey Staff, U.S. Department of Agriculture,[5] discuss the determination, description and classification of *soil profile* properties.

4.1. Some Physical Aspects of Soil Composition

The solid phase of soil may be divided broadly into mineral material and organic matter. The primary source of organic matter in soils is vegetable tissue. Whilst some of the plant organic components are readily decomposed microbially to simple end-products others are converted to relatively stable, but still decomposable, forms and give rise to humus which usually dominates the soil organic fraction.

Humus can be dispersed in water to particles mostly of colloidal dimensions (approximately less than $1\ \mu$), and it is in general intimately associated with the soil mineral material. In terms of weight, humus in non-peaty surface soils is normally less than 10 per cent, and often less than 5 per cent, of the inorganic or mineral material.

The mineral matter itself consists partly of material with a crystal type of structure, which usually predominates, and in part of amorphous or non-crystalline material. Examples of the latter are precipitates of the hydrated oxides of iron and aluminium. Fragments of the parent rock material, ranging from boulders down to about $2\ \mu$ in size, are one source of crystalline minerals.

Rock fragments in this size range have not undergone extensive chemical weathering, which is found to be the distinctive requirement for mineral crystalline material of size below 2 μ approximately.

MECHANICAL ANALYSIS

Under natural conditions most soil particles are clustered together into compound particles or aggregates covering a wide size range, from large clods downwards. Whilst disruption of small aggregates may be quite difficult, such associations are not permanent (section 4.3). A common procedure used in soil classification is to separate it into its individual prime particles, assuming such exist, and then to estimate the percentage by weight in various size ranges or *fractions*. This process, carried out on soil from which "gravel" has been arbitrarily removed by passing it through a 2 mm sieve, and from which organic matter is usually removed, is referred to as *mechanical analysis*. Real accuracy in the mechanical analysis of soils is not possible, partly because the "individual prime particles" of soil depend to some extent on the degree of dispersion and hence on the technique employed. However, real accuracy in mechanical analysis appears unnecessary for agricultural purposes as soil physical properties—let alone plant growth—depend on the nature and not simply on the size of soil particles, and on many other factors as well. This is not to say that the techniques employed in the analysis are unimportant.

Methods of carrying out mechanical analyses are now fairly standardized, and descriptions of the techniques involved are readily available (see Piper,[6] Kilmer and Alexander,[7] and British Standards Institution[8]). A general discussion is given by Baver.[9] The designation and size limits of the various fractions employed in mechanical analysis are standardized. Except in America, where a more detailed classification is usually employed, that adopted by the International Society of Soil Science is used, see Table 4.1. These size limits, whilst arbitrary, at least in some cases correspond roughly to discernible changes in properties, as pointed out by Atterberg, on whose work the divisions are based. For example, below the size limit for clay of $2\,\mu$ there is a marked increase in chemical activity and colloidal behaviour, as already mentioned in connection with weathering. Material in this fraction exhibits characteristic properties such as plasticity when moist, a low permeability to water, and does not feel gritty when moulded between the fingers, as the larger fractions do.

Whilst the coarse sand fraction can be removed by a sieve, the separation of finer fractions is almost universally effected by using their differential terminal velocities of sedimentation in water,

TABLE 4.1

Name of fraction	Size limits expressed as particle diameters
Gravel	Above 2 mm
Coarse sand	2·0–0·2 mm
Fine sand	0·2–0·02 mm
Silt	0·02–0·002 mm
Clay	$< 0{\cdot}002$ mm $(2\,\mu)$

under the action of a gravitational field. If a body falling under the action of its weight (assumed constant) also experiences a resistance to movement which increases in any manner with velocity it will eventually attain a uniform velocity—the *terminal velocity*—often referred to as a *settling velocity* in connection with soil suspensions. This is preceded by a period of acceleration due to resistance being less than the effective weight (i.e. weight minus buoyancy) of the particle in suspension. For sedimenting soil particles, the terminal velocity is quite rapidly achieved, the resistance being due to viscous forces between the settling particle and the fluid through which it is descending. Employing hydrodynamic theory, Stokes (in 1851) derived an expression for the viscous resistance force acting on a spherical body moving with a uniform velocity through an extensive homogeneous fluid, assuming the fluid flow is laminar. The terminal velocity for such a body is given by Stokes' equation:

$$v_t = \frac{g}{18\mu}(\sigma - \rho)d^2, \tag{4.1}$$

where v_t is the terminal velocity (cm sec^{-1}), g is the acceleration due to gravity (about 980 cm sec^{-2}), μ is the coefficient of viscosity of the fluid, $(\sigma - \rho)$ is the difference between the densities of the solid and fluid (g cm^{-3}), and d is the sphere diameter (cm).

Equation (4.1) is applied to soil particles in mechanical analyses to convert settling velocities into equivalent "diameter" of particles (which may be far from spherical, clay particles often being plate-like in form). Considering the purpose of mechanical analysis, such approximations as are involved are not serious. The viscosity coefficient of water decreases about $2\frac{1}{2}$ per cent per °C in the region of 20°C, where its value is about 0·01 poise, and so particle "diameters" are corrected to values which would be obtained at a standard temperature, usually 20°C. The average density of soil minerals is about 2·6 g cm^{-3}. Thus for soil particles settling in water:

$$v = 8·70d^2. \tag{4.2}$$

For soil particles in water, fluid flow is laminar provided particle size is not greater than about 0·25 mm (Keen[10]). The concentration of particles in suspension should be kept lower than about 5 per cent to prevent the flow round one particle affecting that around adjacent particles.

The so-called "pipette method" is the technique thought to be most accurate in determining particle size distributions, and is widely employed. Using a pipette, samples of suspension are withdrawn at a given depth after various periods have elapsed from the commencement of sedimentation. At a depth z below the surface of the suspension of soil particles in water, and at a time t after the commencement of sedimentation (assumed definable), all particles whose settling velocity v is greater than z/t will have passed below this level. All smaller particles will still be descending through the layer above this depth. Let us assume an initial uniform spatial distribution for all particle sizes—a condition aimed at in sedimentation methods. Assume further, as is approximately true for such a system, that particles descend a negligible distance before attaining their terminal velocity. It follows that at depth z and for all time lapses up to t, the concentration of all particles with settling velocities less than z/t is constant, and the same as the initial concentration, since the number of particles entering a layer of infinitesimal thickness at this depth equals the number leaving in any given period. Thus the concentration of particles of such a size that they have settling velocity v is zero above the value of z given by vt, and constant below it. This is the basis of the method. Since in this method z is fixed, v is determined by the time lapse t.

A method employing a hydrometer developed by Bouyoucos[11] is also commonly used. This reflects the variation in suspension density over a finite depth. Though subject to numerous theoretical errors (Black[12]) the hydrometer now in use is adequate and eliminates the drying and weighing of suspended matter necessary in the pipette method. Several other simple techniques involving the change in upthrust on an immersed body are available (Marshall[13] gives an example).

SOIL TEXTURE

As pointed out by Russell,[3] the term *soil texture* should be reserved for use in connection with the mechanical analysis, or for field assessments of the size distribution of particles in the soil. In terms of the three standardized fractions, soil texture depends on the relative proportions of sand, silt and clay. For a given proportion, each of these three fractions is not equally "effective" in influencing soil properties. For example, a given percentage of clay confers greater clay-like properties on a soil than the same percentage of sand or silt. For this reason there is only a general and not necessarily a simple relationship between mechanical analysis and suitable descriptive *textural classes* based on the overall reflection of soil properties as effected by particle size distribution, assuming this can be assessed independently of other factors. Except that the term *loam* is used to describe the textural feature of a soil in which the influence of no one fraction dominates, the terminology of textural classes, such as silty clay or sandy clay loam, is largely self-explanatory. As might be expected, there is no universal agreement on the detailed relationship between textural class and mechanical analysis, though a number of simple diagrammatic means of performing this translation are in common use (Comber[1] or Russell[3] give examples).

The effects of texture on soil properties of agricultural significance are largely indirect, an important example being the influence of texture on the water relations of soil, which will be discussed in Chapter 5.

4.2. The Clay Fraction

The sand and silt fractions of soil are often primarily made up of quartz (the mineral most resistant to decomposition) and primary mineral particles which have undergone little chemical weathering. It is found that the sand fraction, and all but the smaller particles in the silt range, exhibit much less chemical activity than material in the clay size range. For this reason, and also because the clay fraction tends to dominate physical properties to a degree greater than its fraction proportion, this fraction has received much greater attention than the larger fractions. Crystalline material in the clay range is typically differentiated from such material in the larger fractions by being composed of minerals formed as products of complex chemical weathering processes. These *clay minerals* are not found in unweathered parent rocks, and, together with colloidal organic matter (removed in a mechanical analysis) these secondary minerals constitute the bulk of material in the size range of less than $2\,\mu$. It was chiefly the application of X-ray diffraction techniques (described by Cullity[14]) which revealed the mineralized nature of much clay-size inorganic material, and enabled many different types of such minerals to be recognized and their structure determined. A vast literature on this subject now exists and texts such as those by Grim[15] and Marshall[16] describe the clay mineral structures.

Very briefly, such analysis has shown the predominant layer-lattice class of clay minerals to be built up from stacked layers or sheets of oxygen and hydroxyl ions, with metallic cations fitting into the spaces between these layers and bonded to them. Silicon and aluminium are the dominant metallic ions, and silica and alumina sheets interlinked and stacked in different ways is the common feature of the lattice clays. Nevertheless, both silicon and aluminium can be replaced to some extent by other cations when the crystal lattice is in process of formation. Since this substitution depends on *isomorphic* considerations (including ionic size), such substitution is referred to as *isomorphous replacement*. The nature and extent of this replacement gives rise to various members of a clay group with modified properties. It also notably results in the lattice acquiring a net negative electrical charge. Another source of unbalanced charge on clay minerals is due to incomplete charge compensation on terminal atoms of lattice edges (normally oxygen atoms of hydroxyl groups). In the interior of the lattice structure, charge neutralization is complete if there has been no isomorphous replacement. Charge imbalance on the particle edges is thought to be the main cause of net charge on the kaolin group of clay minerals (see p. 96). Whatever their origin, these charges are neutralized normally by relatively low valency cations such as calcium and sodium. These charge-neutralizing cations are therefore not an integral part of the lattice structure, and their place can be taken with varying degrees of ease by other cations. They are therefore known as *exchangeable cations*. Cation exchange is of fundamental importance in soil behaviour, and has received great attention. The process will be briefly discussed later in this section, but a given clay mineral can be found with a variety of exchangeable cations.

Though there now appears a far greater variety of clay minerals than was suspected previously, examples of interstratification and poor crystallization also occurring, Marshall[16] has shown that they may be divided into four main groups. This classification of clays, together with the ratio of the number of silica to alumina layers in the lattice unit (which differentiates two groups of layer-lattice clays), and some information concerning hydration, are given in Table 4.2:

TABLE 4.2

	Group names	Ratio of number of silica to alumina layers	Hydration characteristics
Layer-type lattice (Typical shape is plate-like)	Kaolin or kandite	1:1	Compact sheets, no internal (structural) hydration
	Hydrated mica, or illite	2:1 (non-expanding)	Mostly constant structural layers of water molecules between alumino-silicate sheets
	Montmorillonite	2:1 (expanding lattice)	Expands and contracts markedly with absorption of water between clay sheets
Chain lattice type (Typical shape is rod- or needle-like)	Fibrous clay	(Not appropriate)	Open porous structure; can absorb water internally without expansion

The development of the electron microscope has enabled the shape of clay particles to be directly determined (Baver[9] and Russell[3] contain reproductions of photomicrographs). From such evidence it appears that the majority of clay minerals present in soils are plate-like in shape. Kaolinite, a common clay mineral of the kaolin group, characteristically has well-defined particles of size several orders of magnitude greater than montmorillonitic minerals. Knowledge of the size and shape of such mineral particles helps in understanding and interpreting such macroscopic clay properties as plasticity. The less common chain lattice type of clay minerals have a structure which, in two dimensions, resembles a chain-link fence. The open channels or pores can accommodate both water molecules and exchangeable cations, which are themselves hydrated or surrounded with a water hull.

One reason for the greater physical and chemical importance of the clay fraction lies in the geometrical fact that total surface area per unit mass or volume of material increases with fineness of subdivision. For example, surface area *per unit volume* increases in inverse proportion to the size of the subdivision of this volume for cubical and spherical particles. With such assumptions the specific surface area of $0 \cdot 1 \mu$ particles—in the clay range—would be 200 times that for particles of size $0 \cdot 02$ mm, which is the conventional division between sand and silt fractions in mechanical analysis. Whilst particle shape affects these simple considerations to some extent, more important is the evidence of an "internal" as well as an "external" surface area, at least for the montmorillonite and palygorskite type structures. What is meant by an internal surface area can be illustrated by clays of the montmorillonite group. The bonding forces joining structural sheets of this clay together are weak. Under some conditions adjacent sheets behave as a unit; in other conditions the sheets can separate, thus exhibiting new internal surfaces not available under other conditions. Surface areas of clays are normally measured by absorption techniques, and there is much evidence that whilst water and certain other polar molecules can enter between the internal surfaces of montmorillonite, thus causing the lattice sheets to separate or expand, this internal surface is not available to non-polar molecules. Due to this and other complications, the effective specific surface area of a particular clay mineral, usually expressed in square metres per gram of oven-dry material, is not a unique property. However, in general the surface areas of the three common clay minerals kaolinite, illite and montmorillonite increase in this order, though there can be overlapping in particular cases. The internal surface area of montmorillonite is greater than its external surface, the total specific surface area in general being at least an order of magnitude greater than for kaolinite. Hydration, cation absorption, and other characteristics of high surface area are consequently much more marked in montmorillonites than kaolinites, and the large lattice expansion of the former group of hydration gives rise to marked changes in volume with moisture content.

The factors controlling the type of clay mineral synthesized are not well understood. Rainwater percolating through soil tends to leach out the exchangeable cations, leaving the soil chemically acid (perhaps the most common cause of reduction in crop yields in humid temperate climates). Russell[3] states that as a general rule montmorillonitic clays tend to be found in areas of poor drainage which ensure a good cation supply, and kaolinite clays are usually formed in well-drained areas with a consequent low concentration of bases, with illites forming in intermediate conditions of leaching.

THE CLAY MINERAL–WATER–EXCHANGEABLE CATION SYSTEM

Water is the commonest example of a *polar* type of liquid. The defining feature of a polar molecule is that it tends to take up a unique orientation in the presence of an electric field, and it experiences a moment if constrained to take up any other orientation. In the presence of an electric field polar molecules behave as an electric *dipole*, which consists of two charges of equal magnitude but opposite polarity separated by a finite distance. The *dipole moment* of such a dipole is defined as the maximum moment which it can experience in an electric field of unit strength. The moment is a maximum when the dipole is normal to the field, and zero when aligned parallel to it. For some molecules this moment is a permanent feature, resulting from the molecular structure, and such is the case for water. In other molecules this charge separation is a result of the application of the electric field, and so disappears with the removal of the polarizing field.

For the remainder of this section clay particles will be assumed dispersed in water. Dispersed clay particles are of a size or mass which falls into the arbitrarily chosen range described as *colloid particles*, which are large compared with the normal molecular scale. A clay suspension is a particular example of what in colloid science is referred to as a *sol*. This is defined (Kruyt[17]) as a colloidal system of liquid character, consisting of an intimate mixture of colloidal particles in smaller kinetic units, such as the molecules of a liquid. A dispersion of clay in water, whilst it can exist as such for a very considerable period of time under some conditions, is not in a true equilibrium (or thermodynamically reversible) state, since there is a general tendency for at least some colliding particles to unite. All colloidal particles experience irregular short period changes in momentum, known as *Brownian* motion. This motion is due to the small size of the particles and the thermal motion of the molecules of the fluid in which the colloid particle is immersed. For each infinitesimal time interval there is a resultant momentum communicated to the colloid particle, due to molecular impacts with it, and so its velocity varies very rapidly in direction and magnitude.

For reasons mentioned later, clay particles dispersed in distilled water repel each other. Thus it is only when two approaching clay particles possess sufficient kinetic energy to overcome this mutual repulsion that "short-range" impact takes place. When molecules, or groups of molecules, come into close proximity they attract one another, the forces being of the type considered by van der Waals and London. A van der Waals force is that which gives rise to the constant *a* in van der Waals' modification of the ideal gas equation. It is thus different from the chemical forces arising out of the interpenetration of electron clouds. Van der Waals forces are also distinct from the ionic type of force responsible for the mutual repulsion of negatively charged clay particles, and their properties can be understood on quantum mechanical theory. Van der Waals forces between two isolated atoms have only a very short range of action, of the order of atomic dimensions. However, there is evidence that such forces are "additive", and then it can be shown (Kruyt[17]) that the attractive forces between two colloidal particles would be of much longer range, comparable both with the range of the repulsive forces and with colloidal dimensions. Thus a sol, such as dispersion of clay in water, is in process of gradually "dying" due to the impact of sufficiently energetic clay particles, though in some conditions this may be very slow indeed.

As mentioned previously, a clay particle or *micelle* normally bears a net negative charge. This can be demonstrated by the movement of clay particles towards the positively charged pole in an electric field (*electrophoresis*). In an electrolytic solution this net charge is electrically neutralized by cations, and thus, as at most phase boundaries, there is set up an electrical *double layer*. The term double layer reflects the spatial separation of neutralizing cations (distributed in the solution phase adjacent to the interface) from the negative charge on the surface of the micelle. The most acceptable model of this double layer, due to Stern, is that it consists partly of a layer of ions fixed through absorption at a short distance from the surface of the micelle, and partly of a diffuse distribution of ions throughout a volume adjacent to the micelle. The theory of a diffuse double layer was first given by Gouy and Chapman early in this century, and Stern subsequently modified this model by postulating a fixed absorbed ionic first layer. Because of their irregular Brownian motion, the neutralizing cations in the double layer tend to diffuse away from the micelle surface. However, they are also electrically attracted to the negatively charged micelle, the Coulomb forces being sufficiently "long-range" to affect ions in the diffuse part of the double layer. Under these two opposed influences the cation concentration does not fall abruptly at a particular distance from the surface, but reaches an equilibrium distribution described by an equation due to Boltzmann. This equation shows the ionic density in the diffuse layer as greatest adjacent to the adsorbed layer and gradually decreasing with distance from the surface to the average ionic density in the bulk of the solution. (It is analogous to the decrease of air density with height, due to the action of the earth's gravitational field. The electrostatic field of the charged micelle has an analogous effect on ionic distribution as the gravitational field has on molecular distribution.) Figure 23 illustrates this distribution, and also the effect of the net negative charge on the relative concentration of cations and anions in the vicinity of the micelle.

As a consequence of anion repulsion and cation attraction, anion concentration in solution extracted from the clay is greater than in the solution originally added to the clay, the magnitude of this difference being quite appreciable for solution added to dry clay.

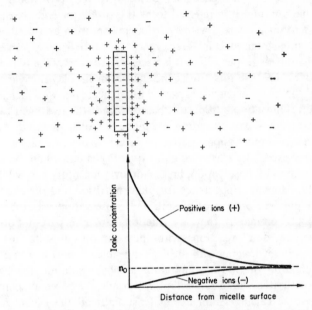

FIG. 23. Illustrating the variation of concentration of positive and negative ions in solution with distance from the surface of a clay micelle bearing net negative charge. n_0 is the ionic concentration in the bulk solution.

This phenomenon is referred to as *negative adsorption* (for a self-explanatory reason). Anions are in effect pushed out from the micellar to the intermicellar solution. The increased loss by leaching of nutrient anions from soils due to this effect may be important.

From the distribution of cations in the double layer (such as is illustrated in Fig. 23), it is obvious that two clay micelles in sufficient proximity for some interpenetration of double layers will experience mutual repulsion. It is this repulsion that has to be overcome before the union of particles is possible.

The total amount of cations adsorbed on soil surfaces per unit mass of soil under empirically defined conditions is approximately independent of the particular cation involved, provided it is a simple ion, and it is referred to as the *cation* (or *base*) *exchange capacity* of the soil, for the particular condition and method of measurement used. It is usually measured at pH7 and expressed in milliequivalents per 100 g. It is necessary to specify the pH of the solution because this affects the charge on the clay micelles, and thus the cation exchange capacity, which decreases with increasing acidity. Since in addition to "base" cations the hydrogen cation always takes part in cation exchange in soils, the term *cation exchange* is more appropriate than *base exchange*, which has been widely used. The reason why this quantity is termed an *exchange* capacity is an observation, first made over a century ago, which is of fundamental importance for plant nutrition and the chemical and physical properties of soil. This observation was that soil, and predominantly the clay fraction, can take part in "double decomposition" reactions with the solutions of salts, the salt cation interchanging with the exchangeable cations. For example, percolation of dilute ammonium sulphate solution through a soil saturated with calcium ions will result in a percolate containing some calcium sulphate, at least until all the exchangeable calcium cations have been replaced by ammonium ions.

Various important chemical and physical processes, such as the weathering of minerals, leaching of electrolytes, soil swelling and shrinkage, and the absorption of nutrients by plants involve ion exchange processes. Since the *exchangeable* (or *counter*) cations serve as the main source of plant nutrient elements, much study has been given to the absorption of cations by plants (Bear[18]). The cation exchange capacity is found to vary greatly with the content and type of clay, and also with organic matter, which, at least in the humidified state, contributes significantly to the colloidal properties of soil. The cation exchange capacity at pH7 (often abbreviated to C_7) is found to range in mineral soils from a few to perhaps 60 m.e. per 100 g (Bear[18]), though the range in well developed humus from mineral soils is usually 150–300 m.e. per 100 g. As might be expected on the basis of specific surface area, C_7 for the main types of clay mineral diminishes rapidly in the order montmorillonite $>$ illite $>$ kaolinite.

We have so far not questioned how the exchange of one cation for another in the double layer takes place, and why one cation should displace, or be displaced by another. Cations are themselves hydrated, the polar water molecules being aligned by the electrostatic field, their effective negative poles being attracted to the positive charge. Nevertheless, a hydrated cation is still very small by colloidal standards and thus undergoes considerable Brownian motion. Thus there must be a continuous exchange or diffusion of ions in both directions between the adsorbed Stern layer, the surrounding micellar solution, and the intermicellar or external solution, although at equilibrium the *net* ionic migration must be zero. On this picture the kinetic exchange of one type of cation for another, both of which are attracted to the negative micelle surface, is not difficult to imagine. It is to such departures from equilibrium, due to the addition of foreign ions, or simply the alteration in concentration of the external solution, that the term *ion exchange* is applied. Although a complete understanding of this exchange process has not been achieved, its general characteristics are recognized. The higher the valency of a cation, the more strongly it will be attracted to a negatively charged micelle surface. Consequently, we would expect, as is found to be the case in general, that the lower the valency of an ion the more easily it is replaced in the exchange process. Another generalization of experimental results (due to Wiegner) suggests that for ions of equal valency, those most highly hydrated are most easily displaced. Thus, although the order is by no means invariable due to the importance of specific effects such as the geometric fit of ions in the clay mineral structure, the ease of *replacement* of the following common soil cations will often be in the order Li $>$ Na $>$ K $>$ Mg $>$ Ca $>$ Ba $>$ Al (their ease of *entry* being in reverse order).

Flocculation and Deflocculation of Clay Suspensions

A suspension of colloidal particles (a sol) which does not settle out under the influence of gravity, or which does so only very slowly, is referred to as *deflocculated* (for reasons which will become apparent). If electrolyte is added to a deflocculated clay suspension, a visible change commences at a fairly definite concentration of electrolyte for any particular clay. The particles of clay begin to come together forming loose aggregates called *flocs*, the suspension as a whole being referred to as *flocculated*. There is a transition between deflocculation and flocculation as electrolyte concentration is gradually increased in which interesting and not well understood phenomena are exhibited (Russell[19]), but for electrolyte concentrations considerably higher than those needed to initiate such changes, flocs begin to form and grow quite rapidly, settling as they do so to form a loose non-uniform sediment, and leaving a clear supernatant liquid above.

The stability of a sol such as a clay dispersion depends largely on the interaction between the particles and on their Brownian motion. If the repulsive energy of the particles (equal to the work which would have to be done to bring them together) is sufficiently great compared with the average energy of Brownian motion, the sol is stable, and the suspension is deflocculated. If this repulsive energy is low, zero, or negative (i.e. there is attraction between the particles), then stability is lost and the system coagulates or flocculates. The production of a repulsive energy barrier corresponds to the concept of deflocculation or *peptization*, and its destruction with that of flocculation, and an important part of colloid science is concerned with an understanding of this energy barrier. Interaction between particles can be considered either in terms of the force between them or their energy of interaction. The force can be obtained from the energy of interaction by differentiating with respect to distance. Analysis is usually in terms of interaction energy, partly because this has to be compared with the energy of the Brownian motion. For a thorough discussion of these processes, reference can be made to Kruyt,[17] Volume 1. Adam[20] (Chapter 8) provides a most useful introduction to the subject.

There must exist an electrical potential difference between the charged clay micelle surface and the intermicellar bulk liquid. This may be considered as made up of the potential drop across the adsorbed Stern layer, and that across the diffuse mobile layer. The potential drop across the diffuse part of the double layer is, to a close approximation, what is known as the *zeta potential* (ζ potential). This potential was originally defined in connection with the relative motion of charged colloidal particles through a suspending fluid when an electrical field is applied, and its magnitude is derived from such migration velocity measurements.

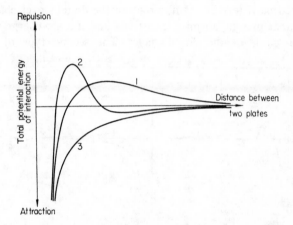

FIG. 24. Illustrating the influence of electrolyte concentration (increasing for curves 1 to 3) on the total energy of interaction between two parallel plates bearing charge of the same sign.

(For this reason it is sometimes referred to as the *electrokinetic potential*.) The ζ potential is closely connected with the stability of all *lyophobic* sols, which are colloidal dispersions of insoluble substances in a liquid medium, such as clay particles in water. Both the magnitude of the ζ potential and the effective thickness of the double layer are reduced by the addition of electrolyte to the solution, and the ionic distribution in the double layer is modified. Since the addition of electrolyte increases the solution conductivity, the electric potential will drop more rapidly with distance from the micelle surface, thus causing the compression in double layers mentioned above. A compression of double layers means that any two micelles will experience repulsion only at a closer approach than before such compression. Thus we have a qualitative explanation of why the addition of electrolytes decreases the stability of clay sols, which is further illuminated by Fig. 24.

The forces between two clay micelles in solution are supposed to be of the two types mentioned earlier. One type, a force of repulsion for particles bearing like charges, is Coulombic in character (i.e. governed by Coulomb's law of interaction between two charged particles). It has its origin in the double-layer charge—both that present on micelle surfaces, and the net charge in the liquid phase of the double layer. The second type of force

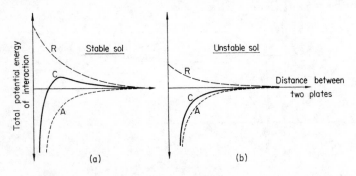

FIG. 25. Showing the combination C of repulsive R and attractive A potential energy curves for two parallel micelle plates bearing charge of the same sign.

between micelles is the general London–van der Waals attraction. With certain assumptions, calculations can be made (Kruyt[17]) of the total potential energy per unit area due to such interaction for two flat parallel charged micelle surfaces. Figures 25a and b illustrate the component and resultant potential energy curves typical of a stable and unstable sol, respectively. In Fig. 25a the combined energy curve exhibits a repulsive energy barrier which has to be overcome before the two micelles can adhere to one another. Stability will decrease and flocculation rate increase markedly when the potential energy barrier falls below the energy of the Brownian motion. Thus the flocculation rate (which may vary from a few seconds to a rate so slow as to be practically immeasurable) is a reflection of sol stability or particle interaction. If a sufficiently large amount of electrolyte has been added for there to be no repulsion (illustrated by Fig. 25b), then the flocculation rate is completely determined by the Brownian motion, since every collision will result in the two particles coalescing irreversibly. The theory of this rapid coagulation has been developed by Von Smoluchowski. Slow coagulation can be interpreted in terms of the smaller number of "short-range" or effective collisions due to a remaining energy barrier.

The transition from the stable to the flocculated condition takes place within a quite small range of electrolyte concentrations. Consequently a *flocculation value* characterizing this transition concentration can be obtained for any particular sol and electrolyte. A most striking result is that for a given sol such flocculation values depend very strongly on the valency of the electrolytic ions of charge opposite to the sol (e.g. on cation valency for a negatively charged clay sol). Furthermore, flocculation values are only secondarily affected by any specific nature of the flocculating ion (other than its valency), or even by the nature of the sol (Kruyt[17]). Thus it is found that for monovalent counter-ions the flocculation values fall between about 25–150 millimols l.$^{-1}$; for divalent ions between 0·5 and 2 millimols l.$^{-1}$; and for trivalent ions of the order of 0·01–0·1 millimols l.$^{-1}$. This long-recognized regularity is known as the rule of Schulze and Hardy. Although there are exceptions to this rule, it is applicable to electrolytic influences on phenomena other than flocculation where such phenomena depend on the electric double layer. Adam[20] presents a simple theoretical explanation of this rule based on the Boltzmann distribution of counter-ions in the double layer solution.

In this section on the clay fractions of soils, we have been mostly considering the structure and colloidal behaviour of "cleaned" pure clay minerals isolated from the soil by appropriate techniques. In the soil such clay particles appear to be covered by some "surface complex" which alters their colloidal properties. For example, profound changes in the colloidal properties of soil can be effected by the addition of organic residues. Thus extrapolation from the behaviour of clay minerals to soil colloids must be made with such reservations in mind.

Despite such reservations, the phenomena of flocculation and deflocculation described for clay sols are of great importance in soils. Calcium is the most common cation in soils. However, if irrigation with saline water has continued for some time, a considerable proportion of the divalent calcium cations may be exchanged for monovalent sodium. From the Schulze–Hardy rule it follows that the clay fraction will remain flocculated only if the water supply has a much higher electrolyte content than would be necessary with calcium saturated soil. A fall of rain, for example, would thus cause deflocculation, and this has serious agronomic consequences, partly due to the filling up of all channels and pores with dispersed clay, which can seriously impede subsequent infiltration of water and exchange of gases between soil and air. The tendency to deflocculate is shown most strongly by clays whose exchangeable ions are predominantly sodium. It is less strongly displayed by potassium, though also monovalent, due to "specific ion" effects.

If the water is removed from a deflocculated clay suspension, a hard uniform deposit is formed with few cracks, if any, in it. On the other hand the sediment of flocculated clay cracks extensively on drying, and is more crumbly and less dense than the dried paste of a deflocculated clay. On re-immersion in water the dried deflocculated clay material with a preponderance of monovalent cations typically redisperses as mentioned above. The agricultural implications of this instability in water will be noted subsequently in connection with soil structure, infiltration behaviour and tillage properties. Such soils form hard large clods when dry, and when wet with rainwater the behaviour of the dispersed clay can be described as liquid mud. This does not occur with calcium saturated clays, and in contrast flocculated clays lead to the formation of crumbs rather than clods in field soils, and these crumbs are more stable on re-wetting than the clods of dispersed soils.

4.3. Soil Structure

The concept of soil structure has been defined in various ways, in broadest terms as "the arrangement of solid particles in the soil profile" (Bradfield[21]). Particular interest is centred in the fact that soil particles can stick together in aggregates, crumbs, and clods of various sizes. Even a compact assembly of clay particles will have a large proportion of its volume available for entry by liquids or gases, as is also the case in another extreme example of soil—a bed of quartz granules such as beach sand. The portion of total soil volume unoccupied by the solid phase is referred to as the *pore space*, and this is increased in soils of agricultural import-ance by the organization of soil particles into aggregates, with the resultant formation of cavities or pores. The structural condition of a soil has a profound influence on water movement (see Chapter 6) and on gaseous exchange—both processes being of great importance to plant growth. Inadequate pore space can also directly hinder root development. Even the emergence of seedlings can be prevented due to a compact poorly structured surface layer, which can be formed by rainfall damage, or by deflocculation of a soil's clay fraction (which can easily occur if sodium is the dominant exchangeable cation). The type and extent of soil aggregation is one factor influencing and influenced by tillage operations. Whilst interest in structural condition is normally centred on the tillage layer, in humid climates stable cracks or fissures in clayey subsoils are most necessary to allow drainage of excess water from the upper layers.

Shattering clods or crumbs, or low-power microscopic examin-ation (Kubiena[22]), reveals that they can be made up of smaller granules which are themselves aggregations of soil particles.

Without suggesting that these smaller granules or micro-aggregates have sufficient discreteness or persistence to be properly regarded as building units of the crumb structure, it is useful to have some physical model on which to understand their construction and properties. The model described below is that proposed by Emerson,[23] which he considers to be consistent with existing knowledge on a range of soil properties. In the terminology of this paragraph, his model refers more to granules or micro-aggregates than to an entire crumb, though nothing like a rigid division is intended. Aluminium hydroxide cementation, a mechanism of crumb formation which may be of some importance in tropical soils, is not included in the model.

EMERSON'S MODEL OF THE MICRO-STRUCTURE

Emerson makes the unrestrictive caveat that the model does not apply to soil in which the uptake of water between the clay plates or crystals is prevented (for example, by heat treatment); nor would it apply to crumbs in which the clay was *pure* kaolinite, because on immersion in water he found that a rearrangement of such crystals leads to the collapse of oriented flakes, no aggregates being formed which possess sufficient stability when wet to be agriculturally significant. Either of these exceptions would be rarely encountered under field conditions. The model is partly based on investigations of the effects of organic matter (Emerson and Dettman[24]) and synthetic soil conditioners (Emerson[25]) on crumb structure, and on the wet strength of crumbs (Dettman and Emerson[26]). It is in some respects a development of a model used by Keen.

The two main constituents of soil being quartz and clay, the problem is how they are bound together to form micro-aggregates. Whilst it is true that in many soil micro-aggregates the clay forms a more or less continuous network enmeshing all larger particles, this is not a very adequate model. Emerson's model is based partly on the concept of a *clay domain*, defined as a group of clay crystals (having suitable exchangeable cations) "which are orientated and sufficiently close together for the group to behave in water as a single unit" (see also Aylmore and Quirk[27]). Emerson assembled evidence that in normal agricultural soils, where calcium is the dominant exchangeable cation, layer-lattice type clay crystals congregate into such oriented domains on drying. However, if the majority of the exchangeable cations are sodium, such water-stable domains may not be formed, especially if the electrolyte level is low, and this might be a more important restriction in application of the model than those given above.

The model expresses the possible ways and (in general terms) the mechanism whereby individual quartz particles can be joined together into micro-aggregates, and is illustrated in Fig. 26. The

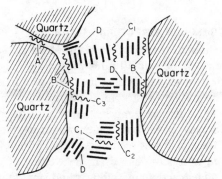

Fig. 26. Possible arrangements of domains, organic matter, and quartz in a soil crumb (after Emerson[23]).

Types of bond:

        A. Quartz–organic matter–quartz.
        B. Quartz–organic matter–domain.
        C. Domain–organic matter–domain.
            C_1, face–face
            C_2, edge–face
            C_3, edge–edge
        D. Domain–domain, edge–face.

model proposes that soil organic matter stabilizes the crumbs primarily by strengthening the bonds between clay domains C, and between the quartz particles and domains B, though the quartz particles may also be directly linked by organic matter A. Emerson also suggests that synthetic soil conditioners fulfil a similar role to the organic matter illustrated in Fig. 26. Under some conditions kaolinite micelles appear to bear a net positive charge on their edges, so that positive edge–negative face attraction is a possibility *D*, though perhaps not common.

Emerson[23] has discussed some applications of this model. Though the model may have to be modified as new information is obtained, and will certainly be filled out in more detail, it is given because the formation of models can assist in relating a variety of properties to simpler and more fundamental concepts. Furthermore, there is often the possibility of progress by investigating previously untested predictions based on such models.

Causes of change in structural condition

These have been discussed by Russell.[3, 28] In tropical and subtropical regions, where rainfall rates can be high and drop sizes large, the effect of raindrops is commonly the most powerful cause of disintegration of structural aggregates into their component particles or into very much smaller aggregates. Important differences in clay behaviour under raindrop action will be illustrated at the end of this section, and similar investigations on soil referred to in section 4.4. Even the wetting of dry crumbs, without the added action of falling raindrops, tends to cause breakdown to an extent which increases with the content of expanding-lattice type clays. The magnitude of the effect thus depends on the initial water content (Panabokke and Quirk[29]), and the rate of wetting (Emerson and Grundy[30]). Water content appears largely to control the structural effects of tillage operations, which may both break down and assist in building up structural aggregates. In the preparation of a seed bed a certain amount of intentional structural breakdown is commonly necessary for good seed germination and emergence, and aggregation usually deteriorates with continuous cultivation (Clarke and Marshall,[31] and Rose[32]), though this is not necessarily or even usually due to the direct effects of cultivation.

Organic matter, particularly in its more stable form known as humus, plays an important role in aggregation, as can be understood from Emerson's model described above. Since microbiological decomposition of vegetable tissue is responsible for the main components of the humus fraction of soil, effective microbial populations would be expected to contribute to good structural conditions. In most situations at least, a grass ley is also observed to have beneficial effects on soil structure for some time after it is removed (Russell[3]). Perhaps enough has been said to indicate that a complex interaction of many factors is responsible for changes in the structural conditions of any soil, and that our present understanding of these factors is still very limited.

Without suggesting that flocculation is a cause of crumb formation in soils, the hardness, size, and water stability of crumbs does depend on whether they contain flocculated or deflocculated clay. Reasons for this, and the effects of addition of water with various levels of electrolyte, were discussed in section 4.2.

ASSESSING THE STRUCTURAL CONDITION OF SOILS

A useful reflection of the state of aggregation of a soil is given by its *moisture characteristic*, which will be described in section 5.8. The slope of this characteristic is proportional to the frequency distribution of pore sizes present, any given pore being drained at a pressure depending chiefly on its size. An important property of a given state of aggregation is its stability to the cycles of wetting and drying it will receive in the field. This can be investigated by following any changes in pore size distribution when crumbs are taken through successive wetting and drying cycles, as was done by Childs.[33]

Descriptions of various methods which have been found suitable for studying long term structural changes in soils under the climatic conditions of Great Britain have been given by Low,[34] together with some typical results. Methods of measuring soil structure have also been reviewed by Russell.[35] Various

measures of the size distribution of structural units in water after
some arbitrary treatment such as wet sieving, have commonly
been employed for assessing the stability of the structure to
wetting, or for comparing such stabilities (e.g. Yoder[36]). The
results obtained are dependent on the technique employed, and,
as would be expected from the above discussion, the dryness of
the soil before wetting and the rate of wetting of the crumbs affects
the results considerably (e.g. Robinson and Page[37]). Thus, on
Emerson's model of aggregation, such methods reflect the
proportion of bonds broken by the rapid swelling of the clay
domains on wetting, and to some extent the stability of remaining
bonds to relatively gentle motion when submerged in water. It
was in an attempt to measure the wet strength of crumbs, unin-
fluenced by the breakdown of dry crumbs on wetting (slaking),
that Emerson[38] introduced the sodium-saturation test. In this
test (later modified by Dettman and Emerson[26]) the gradual
decrease in permeability of soil crumbs when percolated with
$0·5 N$ sodium chloride solution has been used to compare the
cohesion of wet soil crumbs of similar clay content. The permeability decrease is apparently chiefly due to swelling of the clay
which is increasing its hydration, and from the resulting failure of
soil aggregates. These processes are pronounced when sodium
cations are employed, the amount of swelling depending on the
strength of the organic matter bonds. The normality of percolate
was chosen low enough to produce large swelling (and so a big
decrease in permeability) but not so low as to induce noticeable
deflocculation of the clay.

Pereira,[39] after critically examining a number of techniques
developed in temperate climates for assessing changes in the
structural condition of soils, reached the general conclusion that
for the tropical soils tested these techniques were not sufficiently
sensitive to indicate degrees of structural change that are smaller
than those which can be detected by field observations. The rainfall acceptance method of Pereira,[40] in which infiltration into
undisturbed samples under rainfall is measured, has been useful
with tropical soils, as has also the comparison of detachment of

dislodgement of soil under intense laboratory-simulated rainfall (Rose[32]). In both methods the way in which water is supplied to soil in the field is simulated in a simple way.

THE STABILITY OF PURE CLAY CRUMBS UNDER RAINFALL ACTION

The adhesion of clay domains for one another under the action of falling raindrops is clearly of importance to structural stability, as it is also to the maintenance of good rates of water absorption by an exposed soil surface. This has been investigated for artificial crumbs formed from each of three common types of clay material —kaolinite, illite, and montmorillonite (Rose[41]). The cylindrical crumbs (formed by extrusion at a suitable moisture content) were brought to moisture equilibrium with saturated water vapour before testing, thus eliminating the effects of slaking mentioned above. All the clays were saturated with calcium cations, and none dispersed on simple immersion in distilled water. An approximately 1 cm thick bed of loose crumbs was supported on taut open-weave cloth in small cylindrical containers and exposed for half an hour to the moderate simulated rainfall described in the legend to Plate I. The same plate shows the strikingly different behaviour of the clay crumbs to the same rainfall exposure. The kaolinite crumbs rapidly broke down and the kaolin dispersed, in marked contrast to its behaviour *in* water without the mechanical action of the drops. Practically no crumb breakdown and absolutely no dispersion was observed for the clay fraction of Willalooka subsoil, for the montmorillonite, or for a bed of crumbs formed from a mixture of equal weights of kaolinite and Willalooka clay.

After exposure the bed of broken down and dispersed kaolinite had an infiltration rate not greater than 0.05 in. hr^{-1}, whereas for the beds of the other clays the infiltration rate was greater than 100 in. hr^{-1}. Thus for regions where rainfall can be intense, the structural behaviour under raindrop action should be taken into account when assessing the field behaviour of a soil.

4.4. Soil and Water Conservation

Throughout the tropical and subtropical regions, rainfall rates are usually higher than those common in temperate climates. Such high-intensity rainfall produces structural breakdown of the surface layers of soil, impedes infiltration, and can lead to accelerated erosion. The serious significance to agriculture of such direct effects of rainfall has been clarified by the investigations of Laws,[42] Ellison,[43, 44] Hudson,[45, 46] McIntyre[47, 48] and Rose.[49, 32, 41] In particular the vital role of raindrop impact has been established, such impact being responsible for the detachment of soil which can then be transported by run-off water, and for the breakdown of surface structure to form a seal, which can vastly increase run-off (Duley[50]). The more widely this is appreciated in areas where rainfall can be heavy, the less likely it is that the more advanced and disastrous stages of accelerated soil erosion will ever be reached.

Even if the reasons are not always fully recognized, successful systems of indigenous agriculture have often been developed in areas of high erosion hazard. Russell[51] found that one main general aim of crop husbandry and soil management in tropical areas was to keep the soil surface protected from the rain for as long as possible each year. This can be achieved by mixed cropping, the use of suitable rotations, mulching (Jacks et al.[52]), and by leaving crop residues to provide surface protection (Duley and Russell[53]). Denser planting of a crop such as maize which otherwise gives poor soil protection may also be desirable (Hudson[46]).

Rainfall detachment and run-off removal, or run-off scouring and transportation are by far the most common types of water-caused erosion. Bennett[54] describes other types of erosion, and gives a wider description of soil and water conservation problems and practices. Though water-caused erosion is much more extensive on a world-wide basis, erosion by wind can be extremely serious in dry areas, especially with soils possessing only small structural aggregates. Bagnold[55] has shown that the primary process of sand movement due to wind is the cumulative detaching effect of particles removed, accelerated and returned to the surface by the wind. Chepil[56] has investigated soil erosion due to wind.

Soil and water conservation are vitally important considerations in decisions on land-use policies. There is an urgent need for more adequate information on which to base such policies in many parts of the world. A very effective method of obtaining such information has been developed in East Africa by the East African Agricultural and Forestry Research Organization.[57] Like other catchment basin studies, paired catchments were employed. These were chosen as close together and as hydrologically similar as possible. One basin was left unchanged, thus providing a "control" catchment, with whose hydrologic behaviour the other modified catchment was compared. Only rainfall input and streamflow output from such catchments are commonly measured in the conventional approach to such problems, and then long periods (sometimes twenty years!) are necessary to establish conclusions as to the hydrological effects of changes in land use. The extremely important contribution of the co-operative research directed by the E.A.A.F.R.O. team has been to demonstrate that if *all* the main components of the hydrologic balance are determined, then precise information on the effects of such changes is forthcoming in a short period, and most useful information is available within a year or two. Furthermore, the fruitfulness of this intensive approach has been exhibited in situations ranging from high altitude rain forest to overgrazed semi-arid ranchland.

A balanced consideration of the many factors of importance in plant growth may necessitate some compromise with the ideas of soil protection outlined above. For example, one effect of a mulch is to reduce soil temperatures. This can reduce the amount of mineral nitrogen when the soil is wetted, since the amount is a function both of the temperature at which the soil is dried and the period that the soil is in a dry state prior to wetting (Birch[58]). It should also be mentioned that in some climatic situations—for example in the higher altitude tropics—rainfall can play a useful role in breaking down rough-ploughed land, fulfilling a somewhat similar function to winter frost in temperate regions.

Thus an understanding of the reasons for and mechanisms of soil erosion is only one aspect of the many interacting factors that have to be co-ordinated in establishing a continuously productive farming system suitable to a particular soil–topography–climate situation with available resources. As Downes[59] has cogently argued, soil and water conservation are fundamentally an ecological problem of adjusting the system of land use to suit the environment. In raising the productivity of the environment, or in producing plants and grazing animals of more use to man, the balance of the natural environment is inevitably upset. To sustain such greater productivity it is essential that a new equilibrium be found, or the ecological catastrophe of accelerated soil erosion may lead to a serious reduction in productive potential, or to even more disastrous consequences of which we are painfully aware.

Bibliography

1. COMBER, N. W., *An Introduction to the Scientific Study of the Soil*. 4th ed., revised by W. N. Townsend. Arnold, London, 1960.
2. ROBINSON, G. W., *Soils: Their Origin, Constitution and Classification*. 3rd ed. Allen & Unwin, London, 1951.
3. RUSSELL, E. W., *Soil Conditions and Plant Growth*. 9th ed. Longmans, London, 1961.
4. CLARKE, G. R., *The Study of the Soil in the Field*. 4th ed. Clarendon Press, Oxford, 1957.
5. Soil Survey Staff, Bureau of Plant Industry, Soils, and Agricultural Engineering. *Soil Survey Manual*. Handbook No. 18, United States Department of Agriculture, 1951.
6. PIPER, C. S., *Soil and Plant Analyses*. Interscience Publishers, New York, 1944.
7. KILMER, V. J., and ALEXANDER, L. T., Methods of making mechanical analyses of soils, *Soil Sci.* **68**, 15 (1949).
8. British Standards Institution. *Methods of Test for Soil Classification and Compaction*. Waterlow, London, B.S. 1377: 1948.
9. BAVER, L. D., *Soil Physics*. 3rd ed. Wiley, New York, 1956.
10. KEEN, B. A., *The Physical Properties of the Soil*. Longmans, London, 1931.
11. BOUYOUCOS, G. J., An improved type of soil hydrometer, *Soil Sci.* **76**, 377 (1943).
12. BLACK, I. A., Theoretical errors of hydrometer methods, *J. Soil Sci.* **2**, 118 (1951).

13. MARSHALL, T. J., A plummet balance for measuring the size distribution of soil particles, *Aust. J. Appl. Sci.* **7**, 142 (1956).
14. CULLITY, B. D., *Elements of X-Ray Diffraction.* Addison-Wesley, Massachusetts, 1956.
15. GRIM, R. E., *Clay Mineralogy.* McGraw-Hill, New York, 1953.
16. MARSHALL, C. E., *The Colloidal Chemistry of the Silicate Minerals.* Agronomy, Vol. 1. Academic Press, New York, 1949.
17. KRUYT, H. R., *Colloid Science.* Vols. 1 and 2. Elsevier, London, 1949.
18. BEAR, F. E. (ed.), *Chemistry of the Soil.* Reinhold, New York, 1955.
19. RUSSELL, J. L., Studies on thixotropic gelation II—the coagulation of clay suspensions, *Proc. Roy. Soc.* Series A, **154,** 550 (1936).
20. ADAM, N. K., *The Physics and Chemistry of Surfaces.* 3rd ed. Oxford University Press, Oxford, 1941.
21. BRADFIELD, R., Soil structure. *4th Int. Congr. Soil Sci.* **2,** 9 (1950); *J. Soil Sci.* **5,** 57 (1954).
22. KUBIENA, W. L. *Micropedology.* Collegiate Press, Ames, Iowa, 1938.
23. EMERSON, W. W., The structure of soil crumbs, *J. Soil Sci.* **10,** 235 (1959).
24. EMERSON, W. W., and DETTMAN, M. G., The effect of organic matter on crumb structure, *J. Soil Sci.* **10,** 227 (1959).
25. EMERSON, W. W., A comparison between the mode of action of organic matter and synthetic polymers in stabilizing soil crumbs, *J. Agric. Sci.* **47,** 350 (1956).
26. DETTMAN, M. G., and EMERSON, W. W., A modified permeability test for measuring the cohesion of soil crumbs, *J. Soil Sci.* **10,** 215 (1959).
27. AYLMORE, L. A. G., and QUIRK, J. P., Swelling of clay–water systems, *Nature, Lond.* **183,** 1752 (1959).
28. RUSSELL, E. W., Soil structure, *Tech. Commun. Bur. Soil Sci., Harpenden,* **37,** (1938).
29. PANABOKKE, C. R., and QUIRK, J. P., Effect of initial water content on stability of soil aggregates in water, *Soil Sci.* **83,** 185 (1957).
30. EMERSON, W. W., and GRUNDY, G. M. F., The effect of rate of wetting on water uptake and cohesion of soil crumbs, *J. Agric. Sci.* **44,** 249 (1954).
31. CLARKE, G. B., and MARSHALL, T. J., The influence of cultivation on soil structure and its assessment in soils of variable mechanical composition, *J. Coun. Sci. Industr. Res. Aust.* **20,** 164 (1947).
32. ROSE, C. W., Rainfall and soil structure, *Soil Sci.* **91,** 49 (1961).
33. CHILDS, E. C., Stability of clay soils, *Soil Sci.* **53,** 79 (1942).
34. LOW, A. J., The study of soil structure in the field and in the laboratory, *J. Soil Sci.* **5,** 57 (1954).
35. RUSSELL, M. B., Methods of measuring soil structure and aeration, *Soil Sci.* **68,** 25 (1949).
36. YODER, R. E., A direct method of aggregate analysis and a study of the physical nature of erosion losses, *J. Amer. Soc. Agron.* **28,** 337 (1936).
37. ROBINSON, D. O., and PAGE, J. B., Soil aggregate stability, *Proc. Soil Sci. Soc. Amer.* **15,** 25 (1950).
38. EMERSON, W. W., The determination of the stability of soil crumbs, *J. Soil Sci.* **5,** 233 (1954).

39. PEREIRA, H. C., The assessment of structure in tropical soils, *J. Agric. Sci.* **45**, 401 (1955).

40. PEREIRA, H. C., A rainfall test for structure of tropical soils, *J. Soil Sci.* **7**, 68 (1956).

41. ROSE, C. W., Some effects of rainfall, radiant drying, and soil factors on infiltration under rainfall into soils, *J. Soil Sci.* **13**, 286 (1962).

42. LAWS, J. O., Recent studies in raindrops and erosion, *Agric. Engng. St. Joseph, Mich.* **21**, 431 (1940).

43. ELLISON, W. D., Soil erosion studies, *Agric. Engng. St. Joseph, Mich.* **28**: Part I, 145; Part II, 197; Part III, 245; Part IV, 297; Part V, 349 (1947).

44. ELLISON, W. D., Raindrop energy and soil erosion, *Empire J. Exp. Agric.* **20**, 81 (1952).

45. HUDSON, N. W., The design of field experiments on soil erosion, *J. Agric. Engng. Res.* **2**, 56 (1957).

46. HUDSON, N. W., Erosion Control Research, *Rhod. Agric. J.* **54**, 297 (1957).

47. McINTYRE, D. S., Soil splash and the formation of surface crusts by rain drop impact, *Soil Sci.* **85**, 261 (1958).

48. McINTYRE, D. S., Permeability measurements of soil crusts formed by raindrop impact, *Soil Sci.* **85**, 185 (1958).

49. ROSE, C. W., Soil detachment caused by rainfall, *Soil Sci.* **89**, 28 (1960).

50. DULEY, F. L., Surface factors affecting the rate of intake of water by soils, *Proc. Soil Sci. Soc. Amer.* **4**, 60 (1939).

51. RUSSELL, E. W., *Soil Conditions and Plant Growth.* 9th ed. Longmans, London, 1961.

52. JACKS, G. V., BRIND, W. D., and SMITH, R., Mulching, *Tech. Commun. Bur. Soil Sci., Harpenden,* **49**, (1955).

53. DULEY, F. L., and RUSSELL, J. C., Crop residues for protecting row-crop land against run-off and erosion, *Proc. Soil Sci. Soc. Amer.* **6**, 484 (1941).

54. BENNETT, H. H., *Elements of Soil Conservation.* McGraw-Hill, New York, 1947.

55. BAGNOLD, R. A., *The Physics of Blown Sand and Desert Dunes.* Methuen, London, 1941.

56. CHEPIL, W. S., Properties of soil which influence wind erosion, *Soil Sci.* I **69**, 149 (1950); II **69**, 403 (1950); III **71**, 141 (1951); IV **72**, 387 (1951); V **72**, 465 (1951).

57. E.A.A.F.R.O., Hydrological effects of changes in land use in some East African catchment areas, *E. Afri. Agric. For. J.* **27** (entire number) (1962).

58. BIRCH, H. F., Nitrification in soils after different periods of dryness, *Plant and Soil,* **12**, 81 (1960).

59. DOWNES, R. G., The ecology and prevention of soil erosion, *Biogeography and Ecology in Australia* (Monographiae Biologicae Vol. VIII) 472 (1959)

CHAPTER 5

Water and Soil in Equilibrium

BOTH limited availability and over-abundance of soil water directly or indirectly reduce plant growth and crop yields over much of the land surface of the earth. An indirect effect of moisture on plant growth is through its influence on the composition and activity of the soil microbial population. Microorganisms in the soil are continually decomposing fresh plant and animal residues into nitrogeneous and other mineral compounds in forms which can be assimilated by plants. Unless an adequate artificial supply of plant nutrients in an available form is supplied, in the last analysis plant growth depends on this breakdown of organic residues which involves a whole range of organisms including bacteria, fungi, actinomycetes, protozoa and other soil fauna. Though the microflora and fauna of the soil are probably less affected by changes in their microenvironment than higher plants, moisture appears one of the more important factors affecting their number and activity, at least in the case of bacteria (Waksman[1] and Russell[2]). Activity is restricted at both low and—with aerobic bacteria—high moisture contents. Birch[3] has shown how cycles of alternate wetting and drying speed up the oxidation of soil humus and the mineralization of nitrogen in fallow soils in comparison with soil at a constant moisture content, a finding with important agricultural implications. The availability of water also affects the extremely important transformation of atmospheric nitrogen carried out by bacteria which can live symbiotically in "nodules" formed on or in the roots of leguminous and some non-leguminous plants. The less important non-symbiotic fixation by bacteria is also affected.

The more direct effects of water availability on plant processes

121

will be considered in Chapter 8. The subject of water movement in soils will be taken up in the following chapter. The present chapter assumes no net bulk movement of water, so that the water and soil may be considered to be in static equilibrium. It is concerned with defining properties associated with the soil–water system (section 5.1), and with defining and describing the factors which govern the direction in which water tends to move, whatever system it is that contains water. This chapter therefore lays the basis for understanding the movement of water in soils and in plants. This basis consists in examining and defining the total and component *potentials* of water. These definitions can be generalized and extended to form a basis for understanding the purely physical or passive movement of any other translocatable component besides water in the soil–plant system such as plant nutrients.

5.1. The Soil-water System

The mineral and organic compounds of soil form a solid (though not rigid) *matrix*, the interstices of which consist of irregularly shaped pores with a geometry defined by the boundaries of the matrix. That part of the soil not occupied by the matrix is referred to as the *pore space* or *voids*. The pore space is in general filled partly with soil air and water vapour, and partly with liquid phase soil water (Fig. 27).

Consider a particular volume V of soil. This will contain a volume V_s of solids, V_w of water and V_a of air and water vapour so that:

$$V = V_s + V_w + V_a. \tag{5.1}$$

The volume $(V - V_s)$ is the volume of the pore space, and the ratio:

$$\varepsilon = (V - V_s)/V \tag{5.2}$$

is defined as the *porosity*. It is thus the fraction of any total volume of porous material occupied by pores. This aspect of a porous system can alternatively be expressed in terms of a *voids ratio* defined as:

$$\text{Voids ratio} = (V - V_s)/V_s. \tag{5.3}$$

Possible ambiguity in defining the water content of soil is avoided if the water vapour is considered condensed. Nevertheless, the volume of water when it is all in the liquid phase is still almost

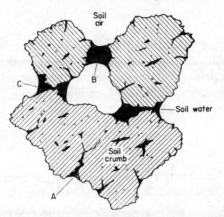

Fig. 27. Soil water within soil crumbs, and suspended between adjacent crumbs.

exactly equal to V_w. Thus the *water content* of soil may be expressed on a *volume* basis (again for a particular volume V of soil) either as

$$\left.\begin{array}{c} \theta = V_w/V \\ \theta_s = V_w/(V - V_s), \end{array}\right\} \tag{5.4}$$

or as

where θ_s is often referred to as the *saturation ratio*. It can also be expressed on a *mass* basis as:

$$w = m_w/m_s, \tag{5.5}$$

where m_w and m_s are the masses of soil water and dry solids respectively. All these ratios may be expressed as percentages.

The mass of "dry" soil m_s is found to decrease gradually as higher oven temperatures are used in drying, water molecules strongly bound to colloids requiring considerable thermal energy for their removal. Water content is calculated in terms of the equilibrium condition at a temperature of 105°C.

Conversion from moisture contents on a mass to a volume basis and *vice versa* requires a determination of the *bulk* (or *apparent*) *density* ρ_b of the soil matrix, defined by:

$$\rho_b = \frac{m_s}{V}. \tag{5.6}$$

Bulk density ρ_b usually increases with depth in a soil profile due to compaction by the soil above, and is best determined by cutting out "undisturbed" cores of known volume from soil in situ, and obtaining the oven-dry mass as described in section 7.1.

From eqns. (5.4), (5.5) and (5.6) it follows that

$$\theta = \frac{w\rho_b}{\rho}, \tag{5.7}$$

where ρ is the density of water (unity in c.g.s. units). From eqn. (5.4) it can be seen that θ can be interpreted as the "equivalent" depth of water in unit depth of soil, this "equivalent" depth being the depth achieved if water were impounded on a horizontal impervious surface. It is thus a more suitable expression for water content than w when considering storage of water by rain or irrigation, or losses by evaporation or drainage. Indeed, as can be seen from eqn. (5.7), using w in such field studies would lead to error unless the variation of ρ_b down the soil profile were taken into account, so that in effect θ *must* be used.

It is common experience that a soil with high clay content may feel dry to the touch and cause plants to wilt while a sandy soil with the same water content may appear quite moist and produce no wilting. This is due to the energy of retention of soil water being greater (per unit mass or volume of water) in the former than in the latter soil. This means more mechanical work would have to be expended in removing a small amount of water from the clay soil than from the sandy soil at the same water content. This is in part what is implied by the term "availability" of soil moisture to plants.

Thus, in addition to water content of soil, it is essential to have information on the energy associated with soil water whether we are interested in the availability of soil water for plant growth, the flow of moisture in soil (as we shall see in Chapter 6), the mechanical properties of soil, or any other soil property. This energy per unit quantity of water in the soil is affected both by the moisture content and by the mechanisms of water retention in the soil, and in some situations the soil water's position in the earth's gravitational field, the gas pressure acting on it, and its chemical constitution also can be of importance. The unit quantity may be unit mass, unit volume, or unit weight. In all cases it is only differences in energy that are significant, and not the absolute value of the energy, even assuming this can be defined. Energy of soil water is a measure of the force fields to which it is subject. Spatial changes in energy per unit quantity of soil water are the cause of movement of such moisture, water tending to move from regions of higher *specific* energy, or energy per unit quantity, to regions where it is lower. Specific energy is a scalar quantity, and the spatial variation of this specific energy of soil water can be described as a "scalar point function"—i.e. a function possessing a definite value, relative to some arbitrary standard state, at each point in the soil. (*Scalar* quantities are those like volume or temperature which involve a magnitude only. *Vector* quantities require direction as well as magnitude for their full specification. Force is an example of a vector quantity.) By definition the work done in transferring unit quantity of water from point A to point B in the soil, under certain restrictions to be given later, is equal to the difference in specific energy of the soil water between A and B. Since work is the product of force with distance moved, the force is equal to the work done in this displacement divided by the distance moved. Thus, in the limit, considering an infinitesimal displacement, the force acting on unit quantity of soil water in a particular direction is obtained by differentiating the specific energy, regarded as a scalar point function, with respect to distance in that particular direction. Thus, denoting the force acting on unit mass of water in any particular direction s by F_s; and the energy per unit mass of water by Ψ:

$$F_s = -\frac{\partial \Psi}{\partial s} \qquad \text{dyne g}^{-1}, \tag{5.8}$$

where the negative sign shows that the force is in the direction of decreasing Ψ.

Any scalar point function (such as Ψ) from which a force or any other useful property can be obtained by differentiation is called a *"potential"* (or a *potential function*). It thus follows from eqn. (5.8) that the specific energy of water has the nature of a potential. This fundamental concept of potential will receive further attention and clarification in the subsequent sections of this chapter.

5.2. The Potential of Water in a Soil or Plant System

The *total potential* (or energy per unit quantity) of water Ψ is defined as the mechanical work required to transfer unit quantity (e.g. unit mass or unit volume) of water from a standard reference state, where Ψ is taken as zero, to the situation where the potential has the defined value. A pool of pure water at an elevation which can be arbitrarily specified, and which experiences a gas pressure of one standard atmosphere has been adopted by the I.S.S.S. (International Society of Soil Science)[4] as this standard reference state. The elevation can be arbitrarily specified, and any gas pressure could be adopted as standard because it is always only *differences* in potential that are significant, and not the *absolute* value of the potential resulting from the definition.

Many forces may affect the total potential of water, but in practice the equilibrium between certain types of forces acting on water (in soil for example) is quite undisturbed in some processes. If so, from the nature of eqn. (5.8) it is unnecessary to consider the contribution of such forces to the total potential, since it remains constant and thus will disappear on differentiation. The contributions to the total potential of soil water that one need take into account in any given application will therefore depend on the type of forces whose equilibrium is affected in the particular process.

In defining the total potential of water in terms of the work necessary to transfer unit quantity (e.g. unit mass or unit volume) of water from a specified reference state, it is assumed that this transfer in no way affects either the potential being defined, or the reference state. However, this is no real problem, since the definition of potential is only used to obtain an expression for the potential in terms of other directly measurable quantities, as will be illustrated in the following sections. The fact that it would be experimentally very difficult, or perhaps even impossible, to measure the work done in the transfer of unit quantity of water involved in defining the potential of any particular state is thus no disadvantage in this rather odd but fundamental type of definition used for potentials. Any attempt to determine potential by measuring the work involved in the transfer of unit quantity of water is made practically impossible because certain restrictions not so far mentioned have to be made as to the manner in which the transfer is to be performed. These restrictions are necessary for the definition of potential to be satisfactory and independent of the particular method used and path taken in transferring the test quantity of water. For such a definition to be adequate the work performed must depend solely on the initial and final state of the water, and not in any way on the particular history of the transfer process. This condition is satisfied by a process which is referred to as *reversible*. Thermodynamic texts (e.g. Pippard[5]) give a fuller description of the limitations necessary for a process to be reversible, but briefly they are that (a) the changes must be performed sufficiently slowly that the substance passes through all stages of equilibrium between the initial and final state; and (b) there must be no frictional forces acting. Such a process can be exactly reversed by an infinitesimal change in external conditions (which is the reason for the name). Besides being reversible in character, such a transfer must also be carried out *isothermally* (i.e. at a constant temperature) if the work done is to be unique.

On first contact with the potential approach to understanding water movement it may appear somewhat theoretically top-heavy compared to other more intuitive approaches which can be made to particular problems. However, an advantage of the total potential approach is that it provides a coherent theoretical framework into which all particular situations and types of problems may be fitted. Once mastered it thus provides a powerful conceptual tool whose usefulness will be illustrated in the remainder of this book. It will be seen in future chapters that such varied subjects as land drainage and the dynamics of osmosis can be understood in terms of the same fundamental concept, namely that water *tends* to move from regions or situations of higher potential to those where it is lower. The application of the concepts of this chapter to water within plants will be discussed in Chapter 8. In the remainder of this chapter it is assumed that water is present in soil. In soil that is uniform in its properties in all directions, bulk movement of water will take place in the particular direction for which F_s in eqn. (5.8) is a maximum. However, structural fissures, compact or porous soil layers may cause conductivity differences in different directions. In soils displaying such *anisotropy*, bulk movement of water does not necessarily coincide with that of the maximum potential gradient (Childs in Luthin[6]).

5.3. Components of the Total Potential of Soil Water

An introduction, to be amplified in subsequent sections, will now be given to the various types of force fields which affect the total potential Ψ of soil water.

Whilst from gravitational theory every body attracts every other body, on the earth's surface it is an adequate approximation for most purposes to neglect all such attractions except that which the earth exerts on every body. Because work has to be done to raise water (or anything else) above the earth's surface against this gravitational attraction, the component of Ψ due to this attraction, called the *gravitational potential Z*, increases with height from the earth's surface.

If the pressure on water in the soil is altered in any way whatsoever, this will also affect its potential, since work would then have to be done in transferring a quantity of water from the reference state defined in section 5.2 to the soil water at a different pressure. This work would be positive if the soil water were at a pressure greater than one standard atmosphere, and negative if less. The pressure in soil water at a particular point will be greater than an atmosphere if that point is submerged beneath a free water surface. The potential due to this cause has not been defined by the I.S.S.S.,[4] but the name *submergence potential*, and symbol *S* is suggested for it.

If the soil is unsaturated the pressure in the soil water is less than that of the local atmosphere. It is convenient to refer to a pressure less than atmospheric as a *suction*. The common childhood discovery that water can be removed from a damp sponge by sucking it is the type of phenomena with which we are here concerned. The application of a suction will also extract water from saturated soil, more water being withdrawn as the suction is increased. Consequently, the greater the magnitude of the applied suction, the lower will be the moisture content of the soil when the soil water has reached equilibrium at that suction. As will be shown in section 5.6 there are two different reasons why the pressure in soil water can be less than atmospheric. In either case the pressure reduction is associated with the location of water close to the soil matrix or within its pores, and the I.S.S.S. (*loc. cit.*) has recommended the two alternative names of *matric or capillary potential M* for the component potential. From their character it follows that submergence and matric or capillary potentials are mutually exclusive possibilities. If either of these potentials is non-zero, the other must be zero.

Another possible cause of pressure change in soil water is a change in the pressure of the air adjacent to it. The pressure of air, or other gas, is commonly altered in laboratory experimentation (section 5.9). The I.S.S.S. decided in their final report not to give a name to the potential due to such external gas pressure changes, pointing out that this component potential is significant only when external gas pressure differs from the standard atmosphere. However, in this text it is convenient to retain the name *pneumatic potential*—used in a draft report by the I.S.S.S. committee on Soil Physics Terminology,[4] and to denote it by the symbol G.

It is also most convenient to go beyond the I.S.S.S. recommendations in using the term *pressure potential P* to refer to the sum of pneumatic and matric/capillary or submergence potentials. Thus

$$P = M \text{ (or } S) + G. \tag{5.9}$$

P sums the effect of all pressure changes on the potential of soil water, whatever the cause of these pressure changes may be.

For the component potentials considered so far it is only of secondary importance that "soil water" contains dissolved solutes obtained from the soil, or perhaps from applied fertilizers. That soil water is not pure water is vital to plant nutrition, and when it is desired to draw attention to this fact, soil water may be referred to as the *soil solution*. Solutes dissolved in water affect its thermodynamic properties and its potential, for example lowering the equilibrium value of its vapour pressure. The effect of solutes on the total potential of soil water becomes of primary significance if the water is separated by a membrane whose permeability to water molecules differs from that to solute molecules. This component potential has been named the *osmotic potential O* by the I.S.S.S., and is of importance in water movement into and through plant roots, in which there are layers of cells which exhibit different permeabilities to solvent and solute. An air–water surface behaves as an ideal semi-permeable membrane, and so this potential is also important in evaporation from soil and vapour movement within soils.

The total potential Ψ of soil water is the sum of all the component potentials mentioned above. Thus

$$\Psi = Z + P + O, \tag{5.10}$$

where pressure potential P is given by eqn. (5.9). Total potential

TABLE 5.1. COMPOSITION ELEVATION AND PRESSURE EXPERIENCED BY THE "REFERENCE POOL" AND THE "SIMILAR POOL" USED IN DEFINING THE COMPONENT POTENTIALS AND TOTAL POTENTIAL OF SOIL WATER

Potential	"Reference pool" Composition	"Reference pool" Elevation	"Reference pool" Gas pressure	"Similar pool" Composition	"Similar pool" Elevation	"Similar pool" Gas pressure
Gravimetric Z	Same as soil solution	Specified	1 Standard Atmos.	Same as soil solution	Same as point in soil where potential is being defined	1 Standard Atmos.
Pneumatic G		Same as point in soil where potential is being defined				Same as on soil water
Osmotic O	Pure water				1 Standard Atmos.	1 Standard Atmos.
Pressure P	Same as soil solution			Test sample is transferred to soil water. Thus "similar pool" not required		
Matric or capillary M and Submergence S	Same as soil solution		Same as on soil water			
Total potential ψ	Pure water	Specified	1 Standard Atmos.			

Ψ has been defined by the I.S.S.S. as "the amount of work that must be done per unit quantity of pure water in order to transport reversibly and isothermally an infinitesimal quantity of water from a pool of pure water at a specified elevation at one standard atmosphere pressure, to the soil water (at the point under consideration)". One may thus imagine a small quantity of water being transferred from a standard reference pool to the soil water, subject to the stated restrictions. As mentioned in section 5.2 the work done does not necessarily have to be measured directly. The restriction to the transfer of an infinitesimal quantity is a formal way of ensuring that the transfer does not affect the reference state or the potential being defined, a problem also discussed in the previous section.

The composition and elevation of the reference pool, and the gas pressure acting on it vary depending on the component potential being defined, as indicated in Table 5.1. It will be found useful to refer to this table in conjunction with the definitions of the component potentials of soil water given in the following sections. As will be shown there, in defining the potentials Z, G and O, the infinitesimal test quantity is transferred from the reference pool not to the soil water, but to another pool containing a solution identical with the soil solution and situated at the same elevation as the point in the soil where the potential is being defined. The properties of this "similar pool" for defining these potentials are also listed in Table 5.1.

The component potentials of soil water will now be discussed in more detail in the order in which they were introduced.

5.4. Gravitational Potential Z

Due to the earth's gravitational field, there is a force of g dynes acting on each gram mass of water, tending to move it vertically downward. This gravitational force is, of course, the weight of the liquid, and it is a basic cause of downward movement of water, whether by downhill surface flow or downward drainage through the soil. If this unit mass is raised (not necessarily vertically) through a height z, work of amount gz has had to be performed on it, where g is the acceleration due to gravity.

Hence the gravitational potential has been defined by the I.S.S.S.[4] as "the amount of work that must be done per unit quantity of pure water in order to transport reversibly and iso-thermally an infinitesimal quantity of water from a pool containing a solution identical in composition to the soil solution at a specified elevation at atmospheric pressure, to a similar pool at the elevation of the point under consideration".

If the point at which the gravitational potential of water is to be defined is at height z above the level of the arbitrarily specified reference pool referred to in the definition, then the work expended in transferring an infinitesimal volume dV of water between the two levels will be:

$$W = \rho(dV)gz.$$

Then the gravitational potential per unit mass $Z_m = W/\rho\,dV$. Thus:

$$Z_m = gz \quad (\text{erg g}^{-1}). \tag{5.11}$$

Z_m is identical with the "potential energy" per unit mass used in mechanics. It is positive or negative depending on whether the location of the point under consideration is above or below the datum level selected, and once that height is known the value of Z_m follows immediately from eqn. (5.11).

It is often more convenient to define potentials on a unit *volume* Z_{vol} or unit *weight* Z_w basis, rather than on a unit mass basis. It follows that:

$$Z_{vol} = W/dV = \rho gz \quad (\text{dyne cm}^{-2}) \tag{5.12}$$

and
$$Z_w = W/\rho g(dV) = z \quad (\text{cm}). \tag{5.13}$$

Z_{vol} has the same nature and units as pressure.

5.5. Submergence Potential S

Consider the component potential of soil water at point A, Fig. 28, submerged at depth h(cm) below the water table, the plane where water pressure is equal to that of the local atmosphere. Let pressure in the soil water measured from the external gas pressure as datum be denoted by p.

Then

$$p = \rho g h \quad (\text{dyne cm}^{-2}).$$

Submergence potential S may be defined in exactly the same words as used by the I.S.S.S. in defining matric or capillary

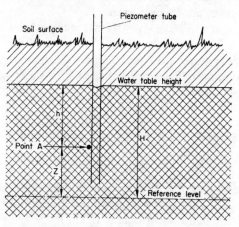

Fig. 28. Illustrating the definition of pore water pressure at a point A distance h below the water table using a piezometer tube. (The significance of height H will be considered in section 5.11.)

potential M, namely as "the amount of work that must be done per unit quantity of pure water in order to transport reversibly and isothermally an infinitesimal quantity of water from a pool containing a solution identical in composition to the soil solution at the elevation and the external gas pressure of the point under consideration, to the soil water". The only difference between the two potentials M and S is that with S, p (the gauge soil water pressure) is greater than zero, and with M, p is less than zero, the absolute pressures thus being greater or less respectively than that of the local atmosphere.

An expression for submergence potential in terms of p will now be obtained from its definition given above. Consider the work required to transfer an infinitesimal quantity of water from the reference pool to the soil water. The stipulation of reversibility ensures no friction in this transfer, which may be assumed to take place through a tube of cross-sectional area dA (cm²) and length l(cm). The work done W in the transfer will then be:

$$W = p\, dA\, l$$
$$= p\, dV \quad \text{(ergs)},$$

where dV(cm³) is the infinitesimal volume of water transferred.

Thus the submergence potential per unit volume,

$$S_{vol} = p \quad \text{(dyne cm}^{-2}\text{)}. \tag{5.14}$$

Also
$$S_m = W/\rho\, dV$$
$$= p/\rho \quad \text{(erg g}^{-1}\text{)} \tag{5.15}$$

and
$$S_w = W/\rho g\, dV$$
$$= p/\rho g$$
$$= h \quad \text{(cm)}, \tag{5.16}$$

where length h is shown in Fig. 28 as the depth of submergence.

Thus the potential S can be determined by inserting a tube or *piezometer* into the soil adjacent to the point where the potential is required (Fig. 28), and measuring the depth h at equilibrium of the point below the free water surface. Length h is sometimes referred to as the *piezometric head* (Terzaghi[7]). An alternative name for submergence potential S could therefore be "piezometric potential".

5.6. Matric or Capillary Potential M

Two mechanisms of water retention by soils will be discussed. Solely from observations on the amount of water removed at various applied suctions, it is impossible to distinguish between these two mechanisms of retention. Consider a saturated soil high in clay content. On the model discussed in section 4.2 this may be considered as a concentrated suspension of small charged clay particles, often plate-like in character. We saw that such suspended micelles usually repelled each other, and that unless they were forced into very close contact, such repulsion increased with the approach of any two particles (Fig. 24). Withdrawal of water from such a soil causes it to shrink, the suction applied to the soil water forcing the particles closer together. For soils high in colloidal material it is found that there is a considerable range of suctions over which this type of response takes place, with the reduction of volume of the soil-water system equalling the volume of water extracted. (Such shrinkage causes damage to buildings and roads when unequal movement of their foundations takes place.)

That this mechanism of water retention *reduces* the energy of free water is shown both by the fact that work has to be expended to withdraw water, and that free water placed in contact with a volume of such soil whose water is at pressure less than atmospheric will flow spontaneously into it. As in all situations water tends to move from where its energy is higher to where it is lower.

Childs[8] and Childs in Luthin[6] have shown that at equilibrium the magnitude of the applied suction must be equal to the mutual repulsion pressure between the charged particles. Whilst the actual hydrostatic pressure varies from point to point depending on the point's microscopic relation to the charged particles, nevertheless the suction of an external body of water in equilibrium with the system is quite definite, and it is this that is taken as the suction in the soil water.

The second mechanism of water retention in soils to be discussed occurs when the volume shrinkage of the soil is less than the volume of water withdrawn, so that air enters the pore space. Except in compact unstructured soils of fairly high clay and water content—conditions which may be found in subsoils—there is usually some degree of air entry into the pore space in desaturation. Water, although completely filling sufficiently small pores (as at *A* in Fig. 27) will then exist typically as necks *B* and annular wedges *C* in the three-dimensional matrix, as is illustrated schematically in cross-section in Fig. 27. Since the pressure in the soil water is less than that in the soil air, which is usually atmospheric, all air-water interfaces must be curved for there to be equilibrium as is shown in any physics text dealing with *surface tension*. Even from the pictorial Fig. 27, it can be seen that the curvatures of air–water interfaces are not simple, partly due to the irregular character of the pore space. An artifice is used to express these curvatures in simple terms. Consider water with an interface in a capillary tube of circular cross-section and radius r. Assuming the contact angle between the capillary tube and water to be zero, the meniscus would be hemispherical and of radius r; and as is well known the pressure in the water on the convex side of the meniscus would be less than the air pressure on the other side of the meniscus by an amount τ, with

$$\tau = \frac{2\sigma}{r}, \tag{5.17}$$

where σ is the surface tension of water. By analogy it is customary to say that at equilibrium under a suction, negative pressure, or pressure deficiency τ, the air-water interface is situated in pores of "effective" radius r, where τ and r are related by eqn. (5.17).

Whether the pressure in soil water is reduced (relative to the external gas pressure) due to either the surface tension effects accompanying air entry, or to the effects of particle repulsion, the associated *matric* or *capillary potential M* has been defined by the I.S.S.S.[4] in exactly the same words as used in the last section to define submergence potential. It therefore follows from the analysis of the same section that:

$$\left.\begin{aligned} M_{vol} &= p, \quad (\text{dyne cm}^{-2}) \\ M_m &= p/\rho \quad (\text{erg g}^{-1}) \\ M_w &= p/\rho g \quad (\text{cm}), \end{aligned}\right\} \tag{5.18}$$

and

where soil water pressure p, measured from the gas pressure acting on the soil water as datum, is now negative.

The concept of this potential was first introduced by Buckingham[9] in 1907, who referred to it as a "capillary potential", and used the symbol Ψ to represent it.

5.7. Determination of Matric or Capillary Potential. Units

This potential is determined by measuring p and using eqn. (5.18). However, p is now negative, and before a simple U-tube manometer can be used to measure it at a particular location in the soil there is a serious difficulty to be overcome. This is how to

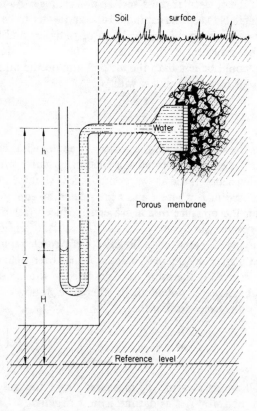

Fig. 29. Porous membrane maintaining continuity between soil water, and water in a manometer, when both are subject to a suction $\tau = \rho g h$. (The significance of height H will be considered in section 5.11.)

maintain the necessary liquid contact between the soil water itself and the external water situated in the manometer, when an attempt is made to impose a suction on the latter so as to achieve equilibrium between the water in the manometer and that in the soil. Water will simply retreat from the open end of the manometer located in the soil, and such an arrangement will indicate zero suction regardless of the suction in the soil water. This difficulty can be overcome by interposing some type of fine-pored membrane between the water in the manometer and that in the soil, continuity between the two existing through the water-filled fine channels in the porous membrane, as is illustrated in Fig. 29. The horizontal lines through the membrane are meant to indicate continuous water-filled pores. Such membranes have been constructed from plates of small sintered glass spheres, unglazed ceramic, and from sheet material such as synthetic sausage casing. In Fig. 29 the manometer arms are shown located in an observation pit, but a more convenient type of instrument, illustrated in Fig. 30, makes this unnecessary.

Pressure p will be given by:

$$p = - \rho_m g h_m + \rho g h_w \quad (\text{dyne cm}^{-2})$$

at the porous membrane in the soil, where the suffix m refers to mercury and w to water. All such instruments for measuring soil water suction are called *tensiometers*. The stopper S shown in Fig. 30 provides a convenient opening for refilling the instrument with air-free water if the suction temporarily becomes so high that air enters from the soil through the fine saturated membrane pores, thus causing a break in liquid continuity. The suction at which such air entry takes place depends on the size of the largest pore. Even with no such air entry breakdown occurs due to entrapped air, or to air coming out of solution at the reduced pressure. Whatever the cause, liquid continuity fails when suctions exceed about 850 cm of water or some 0·8 atmospheres (Marshall[10]), or perhaps lower, depending somewhat on the experimental arrangement. Since soil moisture suctions in the field often exceed this value, other methods described by Haise[11] and Marshall (*loc. cit.*) for example, and in Chapter 7, must then be employed. When the water pressure p is negative, it is found convenient to talk in terms of a positive *matric* or *soil-water suction* τ. Other terms such as soil-water "tension" or "pressure deficiency" have also been used. Thus by definition

$$\tau \equiv -p \quad (p \leqslant 0)$$
$$\equiv 0 \quad (p > 0). \tag{5.19}$$

The matric or soil-water suction has been defined by the I.S.S.S. as "the negative gauge pressure, relative to the external

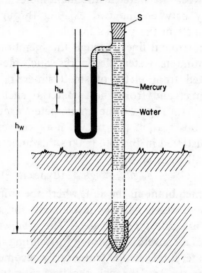

Fig. 30. A form of tensiometer for measuring the suction of soil water *in situ*.

gas pressure on the soil water, to which a solution identical in composition with the soil solution must be subjected in order to be in equilibrium through a porous permeable wall with the water in the soil". Provided the external gas pressure is atmospheric, the water pressure p is equivalent to what is called *pore-water pressure* in the soil mechanics literature (e.g. Terzaghi, *loc. cit.*), in which it is denoted by u or u_w. The I.S.S.S. definitions all assume no pressure or stress in the soil sample due to external loading of the sample. For this reason a quantity called the *suction s* is used in the soil mechanics literature in a way different from the above I.S.S.S. definition of matric or soil-water suction (Croney[12]). I shall refer to this quantity s as the "unloaded suction" for clarity, and it arises as a quantity of importance at depth in compressible (notably "clayey") soils which are unsaturated, but not highly so. The weight of soil and water above any point within the soil results in a total pressure or stress B at that point. This applied pressure in general varies with orientation in the soil. Within the relatively shallow depths of common interest to agriculture, and provided the unloaded suction s is not too low, no great error is introduced into these considerations if stress B is assumed isotropic and calculated from the weight of overburden as shown in eqn. (7.5). These approximations will be assumed in what follows and in section 7.1. They effectively equate B to the *overburden pressure*, which is the total pressure in a horizontal plane in the soil due to the weight of soil above the plane, and due to any other loads borne by the earth's surface. In general this pressure is supported partly by the soil itself, through intergranular contact, and partly by the pore water. Suppose a fraction a of the overburden is borne by the pore water. Then if the soil were *saturated*, we would have $aB = p$. But if it is unsaturated the pore water pressure will be less than aB by some amount which can be determined by removing a small sample from the point in question (in such a way as to leave its structure and moisture content undisturbed) and determining the suction s in its soil water with equipment of the type shown in Fig. 31 of the next section. Then

$$p = aB - s \quad \text{dyne cm}^{-2}. \tag{5.20}$$

The reason for calling s (a positive quantity) the "unloaded suction" is now more obvious. For an incompressible soil, such as sand, $a = 0$ and $p \equiv -\tau = -s$, so that there is no possibility of confusion. For a saturated soil of high clay content a can be unity (Croney and Coleman[13]). At a water table $p = 0$ by definition, and if $a \neq 0$ it follows from eqn. (5.20) that the "unloaded suction" is exactly neutralized there by the component overburden pressure aB. Suppose, for example, the water content of a particular volume of soil is such that its "unloaded suction" s is equal to the pressure exerted by 200 cm of water. Then assuming $a = 1$, the pore-water pressure p (and so the matric suction) will be zero if this soil is located at a depth of 100 cm beneath the surface of a soil whose average wet mass per unit volume is $2\,\mathrm{g\,cm^{-3}}$. In agricultural conditions matric suctions are often high in comparison with aB, in small part due to the coefficient a decreasing as moisture content is reduced, and under these conditions the difference between "unloaded suction" s and matric or capillary suction τ may be negligible.

UNITS

The fundamental units of matric or capillary potential depend on what unit quantity is taken in the potential's definition, these units being given in eqn. (5.18). At pressures where the density of water can be taken $1\,\mathrm{g\,cm^{-3}}$ (a pressure application of 15 atmospheres causing a volume change of only 1 part in 2000), potentials based on unit volume and thus expressed in units of $\mathrm{erg\,cm^{-3}}$ or $\mathrm{dyne\,cm^{-2}}$ are numerically equal to those based on unit mass, with units $\mathrm{erg\,g^{-1}}$.

Writing

$$\tau = \rho g h \quad (\mathrm{dyne\,cm^{-2}}), \tag{5.21}$$

height h can now be interpreted as the height of water column necessary to exert a (positive) pressure equal to the (positive) matric or soil water suction τ. Height h is illustrated in Fig. 29, where it can be seen to represent the suction in terms of the height of an equivalent (hanging) column of water. For soil at equilibrium in the tensiometer shown in Fig. 31 of the next section, h would be equal to $(h_w + \rho_m h_m)$.

From eqns. (5.18), (5.19) and (5.21) it follows that:

$$
\begin{aligned}
M_w &= -\tau/\rho g \\
&= -h \quad (\mathrm{cm}), \\
M_{\mathrm{vol}} &= -\tau \quad (\mathrm{dyne\,cm^{-2}}) \\
&= -\rho g h,
\end{aligned}
\right\} \tag{5.22}
$$

and $\qquad\qquad M_m = -gh \quad (\mathrm{erg\,g^{-1}}).$

Many plants wilt when moisture in the root zone has a value of h of about 15,000 cm. Particularly if it is desired to present information graphically over a wide range of moisture conditions —from saturation to plant wilting conditions, for example—it is convenient to use $\log_{10} h$.

As can be seen from eqn. (5.22), $\log_{10} h$ is a measure of $\log_{10} M$, denoted pF by Schofield,[14] using the analogy with Sørensen's acidity scale (pH) and using F to denote what is referred to as "free energy" in some sections of the physical chemistry literature. Unfortunately, the term "free energy" is taken to denote two different thermodynamic potentials in different literatures. Schofield meant it to imply the thermodynamic potential associated with the name of Gibbs, also called the available energy or net work function, and often represented by the symbol G. I shall hereafter refer to it as the Gibb's potential. However, this identification of Buckingham's capillary potential, now also known as the matric potential M, with the Gibb's potential, whilst it may be a good approximation in many soil situations, is erroneous in principle, since Gibb's potential is affected by changes in other potentials, such as the osmotic potential, to be described in section 5.12, in addition to changes in M. Nevertheless (following Schofield), pF may be identified with $\log_{10} M$, though it is much more commonly taken to mean $\log_{10} h$.

Other units, in which soil water pressure p or suction τ are expressed, are:

1 bar = 10^6 dyne cm^{-2} = pressure exerted by a column of water of height h = 1022 cm if ρ = 1 g cm^{-3} and g = 980 cm sec^{-2}.

1 millibar (mb) = 10^3 dyne cm^{-2} = pressure exerted by 1·022 cm high column of water with the same assumptions as above.

1 (standard) atmosphere = $1·013 \times 10^6$ dyne cm^{-2} (defined in section 3.1) = pressure exerted by a column of water 1035 cm high (with ρ and g as above).

Thus 1 bar = 0·99 atmospheres. Table 5.2 illustrates the relationship between various units at four approximate levels of soil moisture.

TABLE 5.2. VALUES OF MATRIC OR CAPILLARY POTENTIAL M AND MATRIC OR SOIL-WATER SUCTION τ IN VARIOUS UNITS, ILLUSTRATED AT FOUR APPROXIMATE LEVELS OF SOIL MOISTURE

(Adapted from Marshall[10])

| Soil moisture condition | Matric potential M_m or M_{vol} | Matric or soil-water suction | | |
| | | h | h | $\log_{10} h$ |
	ergs g^{-1} or dyne cm^{-2}	dyne cm^{-2}	cm water	atmospheres or bars	h (in cm water)
At suction of 1 cm water (saturation approximately)	-980	980	1	0·001	0
At suction of 100 cm water (roughly approximating field capacity)*	$-9·8 \times 10^4$	$9·8 \times 10^4$	100	0·1	2·0
At suction of 15 atmospheres (approximately wilting of many plants)	$-1·5 \times 10^7$	$1·5 \times 10^7$	15,000	15	4·2
At relative humidity of 0·85 (soils feel dry)	$-2·2 \times 10^8$	$2·2 \times 10^8$	220,000	220	5·4
Conversion from suction h (in cm water)	$-980 h$ (eqn. 5.22)	$980 h$ (eqn. 5.21) Assuming $\rho = 1·0$ g cm^{-3}	h	Atm $= \left(\dfrac{h}{1035}\right)$ bar $= \left(\dfrac{h}{1022}\right)$	$\log_{10} h$

* The meaning of "field capacity" is given in section 6.3.

5.8. Soil Moisture Characteristics

The relationship between water content and matric or soil-water suction is not unique, and depends on the previous history of water intake (*adsorption*) or withdrawal (*desorption*) in a manner to be illustrated. Nevertheless, such relationships are of great significance and utility, and are usually referred to as *soil moisture characteristics*. Tensiometers of the type described in the previous section can be modified to determine such characteristics up to the suction breakdown limit. For the instrument

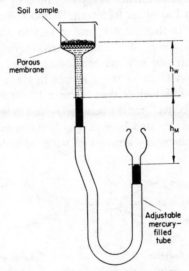

Soil sample

Porous membrane

h_W

h_M

Adjustable mercury-filled tube

FIG. 31. Tensiometer equipment for determining soil moisture characteristics.

illustrated in Fig. 31, the matric suction at equilibrium will be given by:

$$\tau = \rho g h_w + \rho_m g h_m.$$

The mercury can be dispensed with at low suctions when a continuous water column would not be inconveniently long. With such an arrangement some of the soil sample must be removed for determination of water content at each equilibrium suction. Such removals are unnecessary in a modified arrangement where the volume of water moving into or out of the sample can be measured accurately. Also with suitable modifications suctions can be measured without significant exchange of water between the soil and the tensiometer (e.g. Croney, *loc. cit.*).

If the gravitational field is replaced by a centrifugal field, obtained by rotating a suitably modified version of tensiometer in a centrifuge, the suction range can be considerably extended.

Many of the methods available for measuring soil-water suction have been described in detail in a publication by the Department of Scientific and Industrial Research,[15] and some will be briefly mentioned later in this chapter, and in section 7.2.

The general features of the soil moisture characteristics of a *non-shrinking* soil will now be considered. If an increase in the soil-water suction from τ_1 to τ_2 results in the withdrawal of a volume V_{12} of soil water, then volume V_{12} was contained in channels of "effective" radius between r_1 and r_2 where τ_1 and r_1, and τ_2 and r_2 are related by eqn. (5.17). Thus as the suction is increased, the remaining amount of water is reduced, and it is

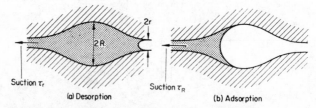

Suction τ_r

(a) Desorption

Suction τ_R

(b) Adsorption

Fig. 32. Cross-section through an idealized void, of maximum effective radius R, interconnected with the rest of the pore space through two channels, of effective radius r.

situated in effectively smaller pores. The pore space may be pictured as irregularly shaped voids interconnected by smaller channels, both voids and channels covering a wide range of sizes. It is a particular feature of such porous systems that causes the equilibrium moisture content at any suction to depend on whether the system is draining or imbibing, a phenomenon referred to as *hysteresis*. This can be understood by considering an idealized void connected by channels (Fig. 32).

The water-filled void shown in Fig. 32a will drain if the suction exceeds τ_r, where $\tau_r = 2\sigma/r$, and the drainage suction is in general determined by the largest effective radius of any connecting channel. However, the criterion governing adsorption by the void is the maximum diameter of the void itself. Considering Fig. 32b, it will fill with water provided the suction is less than $\tau_R (= 2\sigma/R)$, quite independently of channel or neck radii such as r. The implication of these considerations is that at any given matric suction the water content of a soil will be greater on desorption than on adsorption (see Fig. 33a and b).

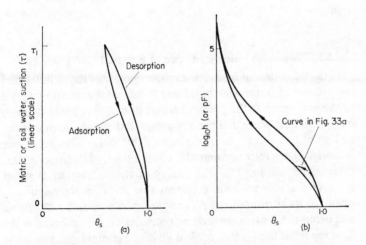

FIG. 33. Soil moisture characteristics showing hysteresis. The arrows indicate direction of change of water content. (a) Both τ and saturation ratio θ_s plotted linearly. (b) τ expressed as $\log_{10} h$, as a function of θ_s.

The relationship between water content and matric suction over a limited range of such suctions is usually referred to as a "scanning curve", with the term "soil moisture characteristic" reserved for the major loop extending up to indefinitely high suction values, as shown in Fig. 33b.

Irreversible rearrangement of soil particles or structural aggregates accompanying moisture changes are another possible cause of hysteresis in soils.

The absolute value of the (negative) rate of change of water content with matric or soil-water suction has been defined by the I.S.S.S. (*loc. cit.*) as the *differential water capacity C*. The *volumetric* differential water capacity would be given by:

$$C_{\text{vol}} = -\frac{\partial \theta}{\partial \tau}, \tag{5.23}$$

and as can be seen from Fig. 33, its value will depend on the particular adsorption or desorption curve appropriate to the situation.

5.9. Pneumatic Potential G due to External Gas Pressure

In the definition of matric or capillary potential the infinitesimal test volume of water was considered to be transferred from a pool to the soil water at the point where its potential is being defined, one restriction being that the gas pressure above the pool and the soil solution should be the same (Table 5.1). The *pneumatic potential* takes into account the effect of any difference in this external "constraint" on the energy of the soil water. It arises because the soil water is a system that is "thermodynamically open" to its surroundings. As mentioned in section 5.3 the name and definition for this component potential was not adopted in the final report of the I.S.S.S. on Soil Physics Terminology, but in the draft report it was defined as "the amount of work that must be done per unit mass (or quantity) of water in order to transport reversibly and isothermally an infinitesimal quantity of water from a pool containing a solution identical in composition to the soil solution at the elevation of the point under consideration at *atmospheric* pressure, to a *similar pool* subject to a gas pressure equal to the gas pressure on the soil water". The atmospheric pressure referred to in this definition is one standard atmosphere (Table 5.1).

Consider the transfer of an infinitesimal volume dV of water referred to in the definition, the volume being transferred from a pool at standard atmospheric pressure A_o to a similar pool subject to a different gas pressure, A_o' say. The suffix refers to conditions external to the soil water. Assuming water to be incompressible up to the maximum value of A_o' applied, this transfer will involve the displacement of an equal volume dV of gas at pressure A_o'. Thus the work done W in the transfer is:

$$W = dV(A_o' - A_o)$$
$$= dV p_0,$$

where p_0 is the external gas pressure measured from standard atmospheric pressure as zero, given by:

$$p_0 = (A_o' - A_o) \quad (\text{dyne cm}^{-2}). \tag{5.24}$$

Thus pneumatic potential based on unit volume is by definition:

$$G_{\text{vol}} = W/dV$$
$$= (A_o' - A_o) \quad (\text{dyne cm}^{-2})$$
$$= p_0 \quad (\text{dyne cm}^{-2}). \tag{5.25}$$

Expressing pressures in terms of equivalent heights of water columns:

$$A_o' = \rho g h_o',$$
$$A_o = \rho g h_o,$$

then

$$G_w = W/\rho g(dV) = (h_o' - h_o) \quad (\text{cm}). \tag{5.26}$$

The relation of both pneumatic and capillary or matric potentials to measurements in the field using tensiometers will now be considered. From eqn. (5.18) $M_{\text{vol}} = p$, where p is measured from the gas pressure acting on the soil water A_o' as datum. Thus, denoting *absolute* pressure on the soil water by p_w:

$$p = (p_w - A_o') \quad (\text{dyne cm}^{-2}). \tag{5.27}$$

However, the datum for gas pressure p_0 in eqn. (5.25) for G_{vol} is one standard atmosphere pressure (A_o, eqn. (5.24)). From eqns. (5.18) and (5.25):

$$G_{\text{vol}} + M_{\text{vol}} = p + p_o,$$
$$= p_w - A_o \quad (\text{dyne cm}^{-2}) \qquad (5.28)$$

from eqns. (5.24) and (5.27).

A tensiometer (Figs. 29 and 30) measures ($p_w - p_a$), where p_a is the local value of atmospheric pressure at the time of measurement. In unsaturated soil, usually $p_a = A_o'$, the pressure in the soil air (Fig. 27) adjacent to the porous tensiometer bulb. However, air may be entrapped in saturated or near-saturated soil, and the difference between A_o' and p_a may then be appreciable.

As mentioned in section 5.2, in practice it is only *differences* in potential that are significant. Hence, provided p_a is essentially *constant*, it follows from eqn. (5.28) that a tensiometer may be regarded as measuring the sum of pneumatic and capillary potentials, though this is strictly only true if $p_a = A_o$. The calculation of both G and M separately from a tensiometer measurement is not possible without further information, but as shown in section 6.1 such separation is normally unnecessary. For such reasons there may be some doubts as to the usefulness of introducing the concept of pneumatic potential in field studies. However, there are important and common laboratory situations where an appreciation of the nature of this potential is necessary, perhaps the most common being the equipment now to be described.

PRESSURE MEMBRANE APPARATUS

From eqns. (5.19) and (5.27):

$$\tau \equiv -p = (A_o' - p_w). \qquad (5.29)$$

In the equipment of Fig. 31 the soil container could be modified so that the air pressure above the soil could be increased above A_o'. Provided p_w is increased equally by raising the fluid level in the adjustable tube, then from eqn. (5.29) τ, and so the soil-water content, will remain unaltered. In particular, if the tube is adjusted so that $p_w = A_0$, then from eqn. (5.29):

$$\tau = (A_0' - A_0), \qquad \text{which from eqn. (5.24)}$$
$$= p_0.$$

Thus the water content of soil is the same when the external gas pressure is atmospheric and the soil water suction is τ, as it is when the soil water pressure is atmospheric and the pneumatic pressure is above atmospheric by an amount numerically equal to τ (Childs, *loc. cit.*). Figure 34 enables the physical reason for this to be seen more readily.

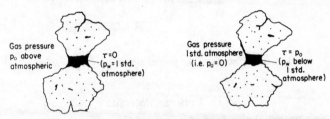

FIG. 34. Soil water at equilibrium under a gas pressure greater than atmospheric, and zero matric suction (left); under atmospheric gas pressure, with a positive matric suction τ (right).

An important advantage of this method of production of soil at equilibrium with a known suction is that it is not limited, as are all tensiometers, by column breakdown at some suction less than atmospheric. An instrument suitable for this method of obtaining soil moisture characteristics is usually referred to as a *pressure membrane apparatus*, and was originally described by Richards[16]. The main features of such equipment are illustrated in Fig. 35, the porous sintered metal plate providing mechanical support for the membrane.

By suitable design, pressures of over 100 atmospheres can be employed if required. After water equilibrium has been achieved at pressure, the apparatus is opened and soil removed from the sample before any appreciable redistribution of water can take place.

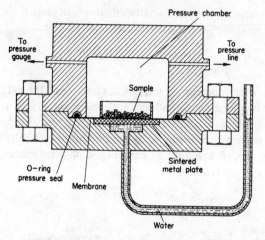

FIG. 35. Principle of the pressure membrane apparatus.

5.10. Pressure Potential P

This potential was defined in eqn. (5.9) as the sum of pneumatic potential and either submergence or matric potential depending on whether or not soil-water pressure p is greater or less than zero.

Pressure potential may be defined (Table 5.1) as the amount of work that must be done per unit quantity of water in order to transport reversibly and isothermally an infinitesimal quantity of water from a pool containing a solution identical in composition to the soil solution at the elevation of the point under consideration at one standard atmosphere pressure, to the soil water.

5.11. Hydraulic Potential Φ

The hydraulic potential is the sum of gravitational and pressure potentials, previously defined. Thus:

$$\left. \begin{aligned} \Phi &= Z + P \\ &= Z + M \text{ (or } S) + G \end{aligned} \right\} \quad (5.30)$$

from eqn. (5.9). It follows from eqns. (5.13), (5.16), (5.22) and (5.26) that Φ_w can be written as:

$$\Phi_w = H \quad \text{(cm)}, \tag{5.31}$$

where
$$H = z + h + (h_o' - h_o) \quad \text{(cm)}. \tag{5.32}$$

The height H is called the *hydraulic head*. In this expression z is the height of the point under consideration above a specified reference level; h is the height of a vertical water column which would exert a pressure at its base numerically equal to the soil-water pressure (when the positive sign is appropriate) or soil-water suction (with the negative sign); $(h_o' - h_o)$ is the difference between the gas pressure on the soil water at the point in question and standard atmospheric pressure, expressed in terms of the equivalent height of a vertical water column. It follows that H may be defined (I.S.S.S., *loc. cit.*) as "the elevation with respect to a specified reference level at which water stands in a piezometer connected to the point in question in the soil". Its definition can be extended to soil above the water table (when the water will be in suction) if the piezometer is replaced by a tensiometer. H is illustrated in Fig. 28 for the soil water of the point labelled A, with water pressure greater than atmospheric; and in Fig. 29 for the soil water of the porous membrane when its pressure is less than atmospheric. In these figures it is assumed that $G = 0$ so that $h_o' = h_o$ (eqn. (5.32)).

The concept of hydraulic head will be employed in Chapter 6.

5.12. Osmotic Potential O

The I.S.S.S. (*loc. cit.*) has defined *osmotic potential* as "the amount of work that must be done per unit quantity of pure water in order to transport reversibly and isothermally an infinitesimal quantity of water from a pool of *pure* water at a specified elevation at atmospheric pressure, to a pool containing a solution identical in composition with the soil solution (at the point under consideration) but in all other respects identical to the reference pool".

A measure of this potential is given by the *osmotic suction* π defined by the I.S.S.S. (*loc. cit.*) as "the negative gauge pressure to which a pool of pure water must be subjected in order to be in equilibrium through a semipermeable (i.e. permeable to water molecules only) membrane with a pool containing a solution identical in composition with the soil water". (This quantity has also been referred to as "solute suction" (Marshall, *loc. cit.*).) A method of measuring π (and so O, see eqn. (5.33) below) will be discussed in section 5.13.

Using similar arguments to that which established $M_{vol} = -\tau$, it follows that:

$$O_{vol} = -\pi \quad (\text{dyne cm}^{-2}). \tag{5.33}$$

The negative sign indicates that if pure water on one side of a semipermeable membrane is transported to a solution on the other side of the membrane, an amount of work $= \pi$ dyne cm^{-2} or erg cm^{-3} is potentially available for the performance of external work as the result of this transport. In the absence of a semipermeable membrane, concentration differences tend to be eliminated as a result of the kinetic activity of molecules, a process referred to as *diffusion*. Although in both osmosis and diffusion there is a tendency to reduce concentration differences, only in osmosis is energy available for the performance of work.

Though osmotic potential affects vapour movement as mentioned in section 5.3, if plant roots and any other differentially permeable membranes are excluded from consideration, this potential has no effect on liquid phase soil-water movement. This is not denying the possibility of chemical redistribution which would take place in the following case for example. Consider two adjacent volumes of soil-water identical in all water potentials except osmotic potential. If these volumes are brought into liquid contact, no net transfer of liquid will take place, but due to diffusion equality of solute concentration and therefore of osmotic potential will tend to be achieved. Osmosis in cells and plants will be discussed in Chapter 8.

A considerable variety in terminology exists in the enormous literature on osmotic transfer. The terms employed in plant physiology have been discussed by Crafts *et al.*[17] Proposals for a unified terminology (Taylor and Slatyer[18]), particularly desirable in plant–soil–water relationship studies, will also be mentioned in Chapter 8.

5.13. Vapour Pressure and the Component Potentials of Water

The magnitudes of the matric M and osmotic O potentials both affect the equilibrium between soil water and its vapour. If either component potential is different from zero, the equilibrium vapour pressure is found to be different from that exerted by vapour in equilibrium with a pool of pure water subject to atmospheric gas pressure at the same temperature. Gas pressure has only a small effect on vapour pressure and is neglected in what follows.

Consider the equilibrium between soil water and water vapour to be reversibly altered by infinitesimal changes in either potential M or O. The equilibrium condition of water vapour can also be represented by a potential Ψ_v. We shall not go into the thermodynamic basis of phase equilibria, which is necessary for a proof of what is presented here simply as an assumption, namely that in the change in equilibrium conditions described above, the infinitesimal change in total potential of soil water $d\Psi$ will be accompanied by an equal change in potential of water vapour $d\Psi_v$. Thus

$$
\begin{aligned}
d\Psi_v &= d\Psi \\
&= dM + dO \\
&= \frac{1}{\rho}(-d\tau - d\pi) \quad (\text{erg g}^{-1})
\end{aligned}
\tag{5.34}
$$

from eqns. (5.18), (5.19) and (5.33) with potentials on a unit mass basis.

An expression for $d\Psi_v$ can be obtained by employing a model. Since we only need to obtain theoretical relationships using the model, it is unimportant that it would be very difficult in practice to maintain the conditions assumed in the model. The model assumes a static constant-temperature atmosphere in complete equilibrium at all heights with a plane surface of pure water located at $z = 0$. Let the vapour pressure at altitudes z and $z + dz$ (Fig. 36) be e and $e + de$ respectively. Just as the total

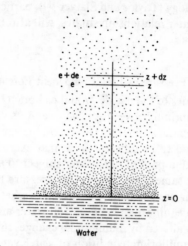

FIG. 36. Section through an isothermal atmosphere in equilibrium with a plane surface of pure water. Dots illustrate the decreasing density of water vapour with height.

atmospheric pressure decreases with height, so there must be a similar decrease in pressure of each component partial pressure, including that of water vapour. At height $z + dz$ the pressure of water vapour will be *less* than that at height z by an amount equal to the pressure exerted by the intervening layer of water vapour, of thickness dz and density ρ_v. Thus de is not a positive quantity (which a differential increment usually represents) and is given, neglecting any variation in g with z, by:

$$de = -\rho_v g\, dz. \tag{5.35}$$

Now $g\, dz$ would be the work done per unit mass in reversibly raising vapour from equilibrium at height z to equilibrium at height $z + dz$ against the gravitational field in this isothermal atmosphere. Thus it is equal to the (positive) increment in vapour potential $d\Psi_v$ corresponding to a (positive) increment in vapour pressure de. Substituting $d\Psi_v$ for $g\, dz$ in eqn. (5.35), and regarding de as a positive quantity (which involves a change of sign) we have:

$$d\Psi_v = \frac{de}{\rho_v}$$

$$= \frac{R_uT}{M_w}\frac{de}{e} \quad (\text{erg g}^{-1}), \qquad (5.36)$$

if ρ_v is replaced with the assumption that water vapour obeys the ideal gas equation (3.1) (a good assumption if no phase change occurs).

Substituting for $d\Psi_v$ from eqn. (5.36) into (5.34) gives:

$$d\Psi = \frac{1}{\rho}(-d\tau - d\pi) = \frac{R_uT}{M_w}\frac{de}{e}.$$

Integrating over a finite change in matric and osmotic suctions from their zero values, and reintroducing suffix M to denote potentials on a unit mass basis, gives:

$$\Psi_m = \frac{-\tau - \pi}{\rho} = \frac{R_uT}{M_w}\ln\left(\frac{e}{e_s}\right) \quad (\text{erg g}^{-1}) \qquad (5.37)$$

where e_s = saturation vapour pressure at temperature T. Water density ρ has been assumed constant over the range of integration. Penman[19] and Hoare[20] give alternative derivations of this result in cases where only one potential is non-zero.

With potential defined on a unit *volume* basis, eqn. (5.37) is:

$$\Psi_{vol} = (-\tau - \pi)$$

$$= \frac{R_uT}{V_w}\ln\left(\frac{e}{e_s}\right) \quad (\text{dyne cm}^{-2}), \qquad (5.38)$$

where $V_w = (M_w/\rho)$ (cm^3 mole^{-1}) is the volume per mole (i.e. molar volume) of water. Strictly, V_w varies a little with solute concentration, and V_w is more accurately the *partial* molar volume $\bar{V}_w$, defined as the change in solution volume per unit change in moles of water, with solvent molarity kept constant. The difference between V_w and $\bar{V}_w$, even with concentrated solutions, is usually only a few per cent.

Equation (5.37) or (5.38) indicates that the sum of the potentials

$(M + O)$ can be interpreted from a measurement of the equilibrium vapour pressure of the soil (or any other material). Neither gravitational nor pneumatic potential can be assigned from vapour pressure measurement, so that Ψ in eqns. (5.37) and (5.38) represents the total potential of water only if these potentials are zero. Particular examples of eqn. (5.37) or (5.38) are commonly used, and so are listed for convenience:

If $\pi = 0$, and using $\tau = \rho g h$ (eqn. (5.21)):

$$h = -\frac{R_u T}{M_w g} \ln \left(\frac{e}{e_s}\right) \quad \text{(cm)}$$

$$= \frac{R_u T}{M_w g} \ln \left(\frac{e_s}{e}\right) \quad \text{(cm)}. \tag{5.39}$$

The negative sign indicates that $e < e_s$ for vapour in equilibrium with a porous medium under suction. Quite large values of h are necessary to make e/e_s appreciably less than unity. For example if $(e/e_s) = 0.99$, substitution into eqn. (5.39) gives $h = 14,000\,\text{cm}$, and for $(e/e_s) = 0.8$, $h = 3.1 \times 10^5\,\text{cm}$. Thus the measurement of matric suction or potential using eqn. (5.39) is practicable only for values of $\log_{10} h > 4.5$ approximately. Because of the rapid variation of vapour pressure with temperature, steady temperatures must be maintained if true equilibrium values are to be measured. Methods of measuring humidity were discussed in section (3.1). The D.S.I.R. publication[15] describes the necessary experimental details for using this method.

Secondly, when $\tau = 0$, such as will be the case if there is a pool of solution:

$$\pi = \frac{R_u T}{V_w} \ln \left(\frac{e_s}{e}\right) \quad \text{(dyne cm}^{-2}), \tag{5.40}$$

where now e = equilibrium vapour pressure of the solution.

Total suction has been defined by the I.S.S.S. (*loc. cit.*) as equal to the sum of matric or soil-water suction and the osmotic suction $(\tau + \pi)$. It is thus the "negative gauge pressure, relative to the external gas pressure on the soil water to which a pool of pure water must be subjected in order to be in equilibrium through a semipermeable membrane with the soil water".

Bibliography

1. WAKSMAN, S. A., *Principles of Soil Microbiology*. Ballière, Tindall & Cox, London, 1927 (p. 583).
2. RUSSELL, E. W., *Soil Conditions and Plant Growth*. 9th ed., Longmans, London, 1961 (p. 216).
3. BIRCH, H. F., The effect of soil drying on humus decomposition and nitrogen availability, *Plant and Soil*, **10**, 9 (1958).
4. International Society of Soil Science, *Soil Physics Terminology*, Bulletin No. 23, 7 (1963). (Draft report No. 20, 2 (1962).)
5. PIPPARD, A. B., *Classical Thermodynamics*. Cambridge University Press, London, 1961 (p. 19).
6. LUTHIN, J. N. (ed.), *Drainage of Agricultural Lands*. Vol. VII of Agronomy. Amer. Soc. of Agronomy, Madison, 1957.
7. TERZAGHI, I., *Theoretical Soil Mechanics*. Chapman & Hall, London, 1951.
8. CHILDS, E. C., The space charge in the Gouy layer between two planes, parallel non-conducting particles, *Trans. Faraday Soc.* **50**, 1356 (1954).
9. BUCKINGHAM, E., *Studies on the Movement of Soil Moisture*, U.S. Dep. Agric., Bur. Soils Bul. **38** (1907).
10. MARSHALL, T. J., *Relations between Water and Soil*. Tech. Comm. No. 50, Commonwealth Agricultural Bureaux, Farnham Royal, 1959.
11. HAISE, H. R., How to measure the moisture in the soil, *U.S. Dep. Agric. Year Book* ("*Water*"), p. 362 (1955).
12. CRONEY, D., The movement and distribution of water in soils, *Geotéchnique*, **3**, 1 (1952).
13. CRONEY, D., and COLEMAN, J. D., Soil moisture suction properties and their bearing on the moisture distribution in soils, *Proc. 3rd Int. Conf. Soil Mech. Found. Eng.* **1**, 13 (1953).
14. SCHOFIELD, R. K., The pF of water in soil, *Trans. 3rd Int. Cong. Soil Sci.* **2**, 37 (1935).
15. Dept. of Scientific and Industrial Research, Road Research Laboratory. *The Suction of Moisture Held in Soil and Other Porous Materials*. Road Research Tech. Paper No. 24, H.M.S.O., London, 1952.
16. RICHARDS, L. A., A pressure-membrane extraction apparatus for soil solution, *Soil Sci.* **51**, 377 (1941).
17. CRAFTS, A. S., CURRIER, H. B., and STOCKING, C. R., *Water in the Physiology of Plants*. Chronica Botanica, Waltham, Mass., 1949.
18. TAYLOR, S. A., and SLATYER, R. O., Proposals for a unified terminology in studies of plant–soil–water relationships, Proc. UNESCO Conf. on Plant Water Relationship in Arid and Semi Arid Conditions (1960). *UNESCO Arid Zone Research*, Vol. XVI.
19. PENMAN, H. L., *Humidity*. Chapman & Hall on behalf of Inst. of Physics, London, 1958.
20. HOARE, F. E., *A Textbook of Thermodynamics*. Edward Arnold, London, 1938.

CHAPTER 6

Movement of Water in Soils

6.1. Fundamental Equations of Liquid Water Movement under Isothermal Conditions

Water in the *liquid* phase in the absence of differentially permeable membranes tends to move in response to gradients in hydraulic potential. It will be assumed that if soil water were subjected to any particular gradient in hydraulic potential, then the flow rate would be the same in any direction in the soil. A soil in which this is so is *hydraulically isotropic.* For such soils there is a considerable body of experimental evidence to show that whatever the state of saturation, the volume flux, or the volume of water crossing unit area per second v, is proportional to the gradient of hydraulic potential in the direction of the flux. Thus, denoting distance in this direction by s:

$$v = -K\frac{\partial \Phi}{\partial s} \quad \text{(cm sec}^{-1}). \tag{6.1}$$

Clearly, v can also be regarded as a *bulk* flow velocity (smaller in magnitude than the fluctuating velocity of microscopic elements of water in their tortuous passage through the pore space). The quantity K, called the *hydraulic conductivity*, is strongly dependent both on the detailed pore geometry of the soil and on its water content. K is thus the volume flux of water resulting from unit gradient in hydraulic potential in the particular soil and water situation under consideration (I.S.S.S.[1]). As mentioned in section 5.2, it is often convenient to express potentials on a unit volume or unit weight basis, instead of a unit mass basis. From eqns. (5.30) and (5.32), where z is positive in the *upward* direction,

$$\Phi_m = gH$$
$$= g[z \pm h + (h_o' - h_o)] \quad \text{(erg g}^{-1}). \tag{6.2}$$

Hydraulic potentials on a unit volume Φ_{vol} and unit weight Φ_w basis are given by:

$$\Phi_{vol} = \rho g H \quad (\text{dyne cm}^{-2}) \tag{6.3}$$

and

$$\Phi_w = H = [z \pm h + (h_o' - h_o)] \quad (\text{cm}). \tag{6.4}$$

In most field situations $(h_o' - h_o)$ is probably small compared with the other components of hydraulic head. With the reservation mentioned in section 5.9, a tensiometer measures the algebraic sum of suction and pneumatic heads $[-h + (h_o' - h_o)]$. The pneumatic pressure term $(h_o' - h_o)$ will be omitted in the remainder of this chapter, so that Φ_w at any point is simply the height H of the water meniscus above the datum level. This meniscus can be located in a piezometer tube (Fig. 28) if water pressure is above external gas pressure. But if the soil water is in suction, a tensiometer must be used, and the meniscus will be in the open limb of its manometer (Fig. 29). Values of H in these two cases are shown in the figures referred to.

Obviously, the dimensions of K in eqn. (6.1) will be different if Φ is taken as Φ_m, Φ_{vol} or Φ_w, the three possibilities being given in Table 6.1.

TABLE 6.1. COMPATIBLE ALTERNATIVE METHODS OF SPECIFYING HYDRAULIC POTENTIAL GRADIENT AND HYDRAULIC CONDUCTIVITY (After I.S.S.S.[1])

Hydraulic potential	Potential gradient	Dimension	Hydraulic conductivity	
			Dimension	Unit
Φ_m	Hydraulic potential gradient (mass basis)	LT^{-2}	T	sec
Φ_{vol}	Hydraulic pressure gradient	$ML^{-2}T^{-2}$	$M^{-1}L^3T$	$g^{-1}\,cm^3\,sec$
Φ_w	Hydraulic head gradient	$LL^{-1} = [0]$	LT^{-1}	$cm\,sec^{-1}$

Experimenting with vertical flow through saturated beds of sand, Darcy (in 1856) was the first to show that volume flow rate was proportional to the gradient in hydraulic head. Generalizations of this result, such as eqn. (6.1), are also associated with Darcy's name. The connection can be traced (Irmay[2]) between Darcy's equation and the fundamental hydrodynamical equation of fluid flow which was derived some 30 years prior to Darcy's work and is known as the Navier–Stokes equation. Such an approach shows that Darcy's equation would be inapplicable if the magnitude of the non-dimensional ratio $(\rho v d/\mu)$ (known as the Reynolds' number) exceeded unity, where v is the "microscopic" flow velocity through channels of size d. This value of Reynolds' number is unlikely to be exceeded under natural gradients of hydraulic potential in soils. A physical implication of this is that the energy made available by the decrease in hydraulic potential accompanying water flow in soils is all used up in overcoming the viscous resistance to that flow.

Consider the vertically *upward* flux of soil water (q_l g cm^{-2} sec^{-1}) with the soil water pressure $p < 0$. The flux is given by:

$$q_l = \rho v \quad (\text{g cm}^{-2} \text{sec}^{-1}). \tag{6.5}$$

The volume flux of velocity v is given by Darcy's equation (6.1). Taking Φ as Φ_w in this equation, with the appropriate K (Table 6.1), and substituting for Φ_w from eqn. (6.4) with the pneumatic term neglected into eqn. (6.1) gives:

$$v = -K\frac{\partial}{\partial z}(z - h)$$

$$= K\frac{\partial h}{\partial z} - K$$

$$= K\frac{\partial h}{\partial \theta}\frac{\partial \theta}{\partial z} - K \quad (\text{cm sec}^{-1}).$$

Thus

$$v = q_l/\rho = -D_l\frac{\partial \theta}{\partial z} - K \quad (\text{cm sec}^{-1}), \tag{6.6}$$

where

$$D_l = -K\frac{\partial h}{\partial \theta}. \tag{6.7}$$

D_l is called the (isothermal) *soil-water diffusivity* (a positive quantity since $\partial h/\partial \theta$ is negative). The magnitude of $dh/d\theta$—the slope of the moisture characteristic—is affected by the consequences of hysteresis. Although K depends very little, if at all, on the direction of water content change, D_l, being the product of both these quantities will not be a unique function of water content. It follows from eqns. (6.7) and (5.20) that

$$
\left.
\begin{aligned}
D_l &= \frac{K}{\rho g C_{\text{vol}}} \quad \text{with hydraulic potential } \Phi_w \\[2mm]
&= \frac{K}{\rho C_{\text{vol}}} \quad \text{with hydraulic potential } \Phi_m \\[2mm]
&= \frac{K}{C_{\text{vol}}} \quad \text{with hydraulic potential } \Phi_{\text{vol}}
\end{aligned}
\right\}
\quad (6.8)
$$

D_l is of fundamental importance in all problems where water content is changing.

The first term on the right-hand side of eqn. (6.6) represents the contribution to water flux caused by any gradient in water content. For horizontal flow gravitational potential is constant, and this leads to the disappearance of the second term on the right hand side of this equation. In this case, "exact" solutions of the flow equation can be obtained for certain boundary conditions and if a class of analytical functions is used to express the relationship between D_l and θ (Philip[3]). (An "exact" solution is one which is obtainable in terms of analytical functions as distinct from solutions that can be obtained only from numerical analysis.) This approach is being developed so that it can be applied more conveniently to practical problems where the relationship between D_l and θ is obtained from experiment (see section 7.2).

Often what is measured experimentally is not q_l but the changes in volumetric water content θ with time at various depths. An expression for $\partial\theta/\partial t$ can be obtained by differentiating eqn. (6.6) with respect to z, giving

$$
\frac{\partial v}{\partial z} = -\frac{\partial}{\partial z}\left(D_l \frac{\partial \theta}{\partial z}\right) - \frac{\partial K}{\partial z}. \quad (6.9)
$$

$\partial v/\partial z$ can be interpreted in terms of $\partial \theta/\partial z$ from conservation considerations. Liquid conservation requires the change in water content of any elementary volume to equal the net flux of water across the boundaries of this volume. Using an approach some-what similar to that employed for heat conservation in thermal conduction (section 2.1), this continuity requirement can be expressed by:

$$\frac{\partial \theta}{\partial t} = -\frac{\partial v}{\partial z}.$$ (6.10)

From eqns. (6.9) and (6.10):

$$\frac{\partial \theta}{\partial t} = \frac{\partial}{\partial z}\left(D_l \frac{\partial \theta}{\partial z}\right) + \frac{\partial K}{\partial z}.$$ (6.11)

Analysis of vertical entry or drainage of water in unsaturated soil, described by eqn. (6.11), has been possible mathematically only through numerical methods (Klute[4] and Philip[5]). Whilst both the experimental difficulties in determining the relationship between D_l and θ, and the mathematical difficulties of analysis are considerable, and complicated by hysteresis effects, the main features of water movement in unsaturated soil can be understood and predicted. This discussion will be taken further in the re-mainder of this chapter.

Flow in saturated soil will be briefly discussed in section 6.4, and the effects of non-isothermal conditions in section 6.5.

6.2. Hydraulic Conductivity

Even assuming it could be adequately measured, the geometry of any soil's pore space, which provides the passages for water movement, is so complex that a calculation of K in terms of this information would be out of the question. The theoretical approaches made have chiefly been to use a much simplified model of the pore space, and to infer the distribution of pores of different size from the moisture characteristic. Such inference is not possible for a soil that shrinks on water removal. Shrinkage is a characteristic of soils with high colloidal content, such soils also being typically compressible. However, with aggregated soil, the interaggregate pore space of even a heavy clay soil can be drained with little shrinkage, and particularly if the structure provides a continuous channel of such large pores these will contribute overwhelmingly to the conductivity of this soil type. In these early stages of water removal from such shrinking soils, and in practically all stages of removal from a non-shrinking soil, an "effective" pore size distribution can be inferred from the moisture characteristic, with the limitations mentioned in section 5.6.

It is implicit in Darcy's law (Philip[6]), that the work done on water moving under a potential gradient is dissipated by viscous shearing stresses. As shown in section 2.2, viscous shear stresses increase with proximity to a surface across which a fluid is flowing, and for this reason the same pore cross-section provides a much more effective fluid conductor if it consists of large pores rather than a greater number of small pores. Thus porosity ε is quite inadequate in itself as an indicator of the hydraulic conductivity of soils.

Poiseuille's equation (Lamb,[7] p. 585) gives the steady volume flow per second through a straight pipe of uniform circular section (assuming laminar flow) to be:

$$-\frac{\pi r^4}{8\mu}\frac{\mathrm{d}\Phi}{\mathrm{d}s}, \tag{6.12}$$

where r = radius of tube,

μ = viscosity coefficient, and

$\mathrm{d}\Phi/\mathrm{d}s$ = gradient in hydraulic potential in the direction s of the tube axis.

Consider a saturated model "soil" consisting of a parallel array of such capillaries, of number n per unit area normal to their axes. Then the bulk flow velocity will be:

$$v = -\frac{n\pi r^4}{8\mu}\frac{\mathrm{d}\Phi}{\mathrm{d}s} \quad \mathrm{cm\,sec^{-1}}$$

$$= -\frac{\varepsilon r^2}{8\mu}\frac{\mathrm{d}\Phi}{\mathrm{d}s} \quad \mathrm{cm\,sec^{-1}}, \tag{6.13}$$

where $\varepsilon = n\pi r^2$, the fraction of unit area occupied by capillaries. This fractional pore area in any plane ε cm^2 cm^{-2} for this model must also be equal to the porosity ε cm^3 cm^{-3} defined in section 5.1. (This is also true in soils with a random distribution of pore space.)

Childs and Collis-George[8] considered the hydraulic conductivity of a model soil consisting of short equal lengths of capillary tube, the radii of which are distributed in a manner inferred from the moisture characteristic. All such capillaries were considered to be randomly in sequence, and the resistance to flow of the larger pore in such a sequence was neglected in comparison with that of the smaller pore—a reasonable assumption since r appears raised to the fourth power in eqn. (6.12). The hydraulic conductivity corresponding to any degree of unsaturation can be considered by taking all pores of effective radius less than that drained at the appropriate soil-water suction, and completely neglecting the contribution which such pores could make if filled with water instead of air. Following this approach, good agreement between measured and calculated conductivities have been obtained on non-shrinking porous materials (Childs and Collis-George, *loc. cit.*). Using a very similar model, Marshall[9] has developed a much more convenient method of performing such calculations.

Due to the difficulty of removing samples of field soil without disturbance, and supporting them so that the laboratory-measured hydraulic conductivity is the same as it would have been for the same sample in the field, field measurements of this property should be made when possible. Methods of measuring hydraulic conductivity in the field are discussed in section 7.2.

Laboratory determinations of hydraulic conductivity have their place, especially in testing theory and in investigating how the degree of unsaturation, or other varied conditions, affects conductivity. The practice of using a different name—often "capillary conductivity"—for the hydraulic conductivity of unsaturated soil is unnecessary, and has not been recommended by the I.S.S.S. (*loc. cit.*). If flow in uniform unsaturated soil is due to a gradient in matric suction, there must be an accompanying gradient in water content and thus in hydraulic conductivity. Thus if the hydraulic conductivity at any particular suction of water content is to be obtained, difference measurements must be kept sufficiently small for the gradient at that particular water content to be inferred from them to an acceptable accuracy. Figure 37 illustrates the distribution of parameters connected with the steady flow of water through unsaturated soil. From eqn. (6.1), for v to be constant the product $(-K\,d\Phi/dl)$ must be constant (Fig. 37). Methods for obtaining the relation between

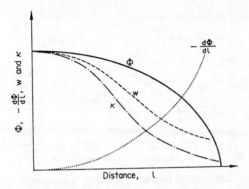

FIG. 37. The type of variation of hydraulic potential Φ, potential gradient ($d\Phi/dl$), water content w and hydraulic conductivity K with distance through a region of unsaturated soil carrying a constant flux of water in the direction of increasing l.

K and water content for laboratory columns have been described by Richards and Weeks,[10] Peck[11] and Youngs.[12]

For reasons discussed in section 4.2, the concentration and composition of dissolved salts in water, particularly in relation to the nature and amount of the exchangeable cations associated with the clay fraction of the soil, can have a most pronounced effect on hydraulic conductivity through the dispersion and swelling which may occur. The interaction between electrolyte concentration, exchangeable sodium content in the soil, and permeability is considerably modified by the structural condition of the soil (cf. section 4.3), and the organic matter content. However, the general type of relationship is indicated in Fig. 38, which Quirk and Schofield[13] considered applicable to most semi-arid soils. For combinations of exchangeable sodium on the soil and electrolyte concentration of applied water above the line shown in the figure, permeability (or hydraulic conductivity) will decrease. The figure demonstrates that to maintain permeability, to each level of exchangeable sodium present in soil there is a lower limit to the electrolyte concentration in applied water.

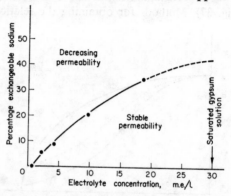

Fig. 38. Relationship between electrolyte concentration in infiltrating water and percentage exchangeable sodium in the soil if permeability is to be maintained (after Quirk and Schofield[13]).

6.3. Infiltration of Water into Soils

Infiltration is the entry into soil of water through its soil–atmosphere interface. The distribution of water content with depth in the soil, referred to as the *moisture profile*, then possesses the features described by Bodman and Coleman[14] and shown in Fig. 39. The saturation and transition zone shown is usually present in soils, but is not observed with porous materials of uniform particle size in experimental situations where there is little possibility of air being trapped under the downward advancing wetting front (Youngs[15]). Water moves through the transmission zone of almost constant water content to the wetting zone where this changes markedly with both depth and time. Figure 40 shows the typical development with time of the moisture profile. Both Figs. 39 and 40 refer to a uniform porous material.

In a series of papers Philip[16–23] has shown how the features of

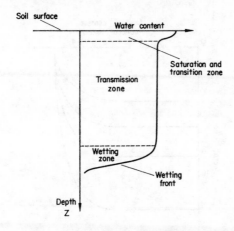

Fig. 39. Zones of the moisture profile during infiltration as described by Bodman and Coleman.[19]

the infiltration moisture profile illustrated in Figs. 39 and 40 (with the exception of the near-surface saturation and transition zone of Fig. 39) can be understood. Furthermore if the dependence of soil-water suction τ and hydraulic conductivity K on water content θ is known, and is assumed unique, the progress of the moisture profile into uniform incompressible soil can be quantitatively determined using eqn. (6.11) provided further assumptions are made. Two further assumptions are that the infiltrating water has no effect on soil structure, either at the surface or in the bulk of the soil, and that the effect of entrapped air on infiltration is negligible. The assumption of a unique dependence of K and τ on θ appears justified whilst the soil surface is maintained at saturation, since only the moisture characteristic for "wetting" is involved. Youngs (*loc. cit.*) working with packed columns of inert isotropic porous materials in the laboratory, found good agreement between experimental moisture profiles and those calculated from this theory. Nielson *et al.*[24] provided a similar comparison on two soils in the field which possessed fairly uniform properties to the depths of wetting investigated. Such lack of agreement as

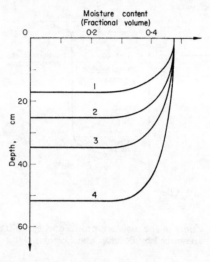

Fig. 40. The development with time of a moisture profile during vertical infiltration from a surface maintained at saturation into an initially dry column of slate dust after (1) 18 min; (2) 39 min; (3) 78 min; (4) 159 min (after Youngs[15]).

was found between measured and calculated profiles may have been due to the assumptions of the theory not being satisfied.

Parr and Bertrand[25] have summarized the main predictions of Philip's theory. For "small" times after the surface application of water to initially dry soil, gradients in soil-water suction are much greater than that due to gravity, and absorption is similar to that which would take place into a horizontal column. The depth of the wetting front then increases as $t^{\frac{1}{2}}$ (t = time), the infiltration rate or water flux at the surface decreasing as $t^{-\frac{1}{2}}$ from very high initial values. After a "long" time the soil becomes saturated, or at least takes a high constant water content, for an appreciable depth below the surface. The flux across the surface must therefore ultimately approach that which is transmitted by the saturated soil under the action of the earth's gravitational potential alone. The velocity of the wetting front also becomes constant. In the unsaturated wetting zone, decreased hydraulic conductivity is compensated by large gradients in matric or capillary potential due to the rapid change in water content. Long after the commencement of infiltration the initial soil-water content has little effect on infiltration rate. However, a higher initial soil-water content initially decreases infiltration rate but increases the velocity of advance of the wetting front (the two results superficially appearing contradictory). Philip's analysis leads to an algebraic infiltration equation, which was compared by Watson[26] with others in use in field studies.

Understanding the redistribution of water in soil *after* it ceases to be available at the surface is even more complex than that described above, chiefly because hysteresis in the moisture characteristic is of vital importance (Youngs[27, 28]). After being saturated to an appreciable depth and then allowed to drain freely, the water content of structured soils often drops fairly quickly, and then remains reasonably constant for periods of a day or so provided surface evaporation is prevented. The water content at this particular drainage stage is referred to as the *field capacity*. (Russell,[29] p. 381, reserves this term for the condition of the soil rather than the water content.) Russell ascribes this common feature to the rapid reduction in hydraulic conductivity following the removal of water from larger interaggregate pores. Though vaguely defined, and depending to some extent on the boundary conditions in its determination, the rapid drainage of water in excess of field capacity makes this quantity a useful upper limit to the water reserves available for plant growth.

6.4. Field Drainage of Soils

In agriculture, field drainage refers both to the steps taken to remove soil water present in quantity exceeding that which is desirable, and to the mechanism of this water removal (Childs[30]). Obviously, drainage becomes important in situations where high soil water contents can persist, but drainage is often an essential requirement in continuous irrigation practice. The continual evaporation of irrigation water can eventually lead to the accumulation of salt in concentrations sufficient to restrict or even prevent plant growth, unless the excess salt is flushed away by the addition of water in excess of that evaporated. This "flushing" of salts can usually only be continued successfully by installing underground drainage channels, as may be provided by an interconnecting system of tile drains.

The need for drainage may be biological or physical. Microbial activity, for example, depends on adequate aeration. Drainage may be desired due to the low mechanical strength of wet soil, which can hinder mechanical cultivation, or lead to undesirable structural effects if such cultivation is performed.

A comprehensive monograph on the drainage of agricultural lands is readily available (Luthin[30a]), and only the briefest outline is given here of the principles concerned with the design of drainage systems. The *water table* is defined as the location where the pore water pressure is zero (i.e. equal to that of the local atmosphere in terms of absolute pressure). Since water pressure will be positive below the water table, any cavity created there will fill with water. In a drainage system this inflow is conducted away, thus maintaining zero water pressure at the drain wall. The problem is to design the drainage system so that the required soil-water conditions will be maintained in a particular soil under given weather and boundary conditions. A much-studied drainage problem is illustrated in idealized form in Fig. 41. In the figure arrows show approximately the flow streamlines from the soil surface to the drains. It can be seen that the problem is defined by known or imposed values of water flux or hydraulic potential at the boundaries such as PQ, QR, RS and SP, and at the drains D. However, the location of the water table boundary, such as PQ, is unknown initially, and its determination is an important aim in solving the problem.

With the assumption that flow is vertically downward in the unsaturated region of soil above the water table, the flux across the water table is then known and equal to the flux across the surface. Solution can then be sought separately for the saturated

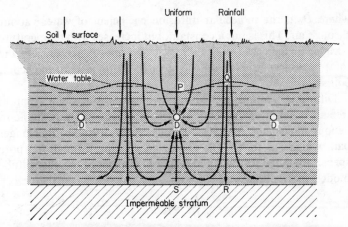

Uniform   Rainfall

Soil ‖ surface

Water table

P

Q

D

D

D

S   R

Impermeable stratum

FIG. 41. A steady-state drainage situation with a water table maintained by steady uniform rainfall in the presence of a drainage system.

below water table region, and the unsaturated region above it. Childs[31] exploited the identity of form between Darcy's law and Ohm's law to obtain solutions to a wide range of drainage problems.

6.5. Water Vapour Movement in Soil

Whilst wind-induced bulk movement and mixing of air in the near-surface pore space can be a cause of vapour movement, this section will be restricted to vapour diffusion, where such bulk mixing is absent. Diffusion refers to the net transfer of molecules of any particular type in response to a gradient in density or partial pressure of that particular type of molecule, provided the net transfer is due to the thermal motion of the molecules, and not due to bulk mixing or stirring, for example. The flux of water vapour in a particular direction through air free of any convectional motion q_v is related to the gradient of vapour density in that direction (z say) by:

$$q_v = -D_o \frac{\partial \rho_v}{\partial z} \quad (\text{g cm}^{-2}\text{sec}^{-1}), \qquad (6.14)$$

where D_o is the molecular diffusion coefficient of water vapour through air. This is not a constant, but increases with temperature because of increased molecular velocity, and decreases with the total pressure because of the greater frequency of molecular collision.

Diffusion in a porous medium such as soil is less rapid than in free air because of the reduced area of cross-section available to vapour movement, and also because of the tortuous and longer path which has to be followed in progressing through the pore space. Marshall,[32] using a random-sequenced capillary tube model, showed that:

$$\frac{D}{D_o} = \varepsilon_g^{3/2},$$

where D is the diffusion coefficient in the porous system of gas-filled porosity ε_g ($= \varepsilon - \theta$). Currie[33, 34] has shown that, in general, a more complex expression must be used for this ratio, and that in a soil–water system the type of variation of D with saturation differed, depending on whether drainage of the larger (chiefly interaggregate) pore space was taking place, or whether water was extracted from the smaller pores within aggregates. This complex relationship will be represented for brevity by:

$$D = D_o f(\varepsilon_g), \qquad (6.15)$$

though the unspecified function $f(\varepsilon_g)$ depends on more than just gas-filled porosity.

Vapour density and pressure gradients in soil can be due to a gradient in temperature (Fig. 42), soil-water suction (eqn. 5.39), or dissolved salts (eqn. (5.40)). Of these three, temperature gradients are in general the most important. If theory is developed along these lines (Philip and de Vries[35]) it is found to predict vapour fluxes lower than those experimentally determined, differences of an order of magnitude being found in some cases. Philip and de Vries have suggested that this discrepancy is due to two factors neglected in the simple approach, and of importance at moisture contents below that at which water forms a continuous network

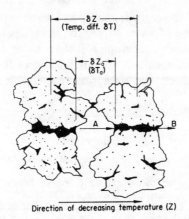

Direction of decreasing temperature (Z)

FIG. 42. Illustrating the modifications to simple diffusion of water vapour in soil where liquid continuity has failed.

throughout the soil. When liquid continuity has failed, water exists as isolated "necks" and "islands" illustrated in Fig. 27. Vapour flow then takes place through the gas-filled cavity (in the direction of decreasing temperature, for example), but it does not have to avoid a water-filled neck as was assumed in the simple theory. It can proceed by condensing on the "upstream" side of the neck and re-evaporating on the "downstream" side, as at A and B in Fig. 42. Thus the *entire* pore space ε, and not just ε_g, is effective in vapour diffusion under such moisture conditions. D. Rose (Ph.D. thesis, London University) has found that such liquid–vapour interaction takes place under isothermal conditions, and is not restricted to temperature-induced diffusion as Philip and de Vries suggested.

The second modification arises from the temperature gradient across air-filled pores (arising from temperature difference δT_a, Fig. 42) being greater than mean temperature gradient in the medium as a whole (calculated from δT, Fig. 42). The combined correction factor η, of the form:

$$\eta = \frac{\varepsilon_g + \theta}{f(\varepsilon_g)} \cdot \frac{(\delta T_a/\delta z_a)}{(\delta T/\delta z)}$$

has been found to reconcile theory and some experimental data.

The proportion of water movement as vapour to that as liquid will clearly rise with decreasing soil-water content. It is therefore under such conditions that vapour movement is likely to be of greatest importance, but its agricultural significance is not yet clear.

Bibliography

1. International Society of Soil Science, Soil Physics Terminology, *Bulletin No. 22*, 5 (1963).
2. IRMAY, S., On the theoretical derivation of Darcy and Forshheimer formulae, *Trans. Amer. Geophys. Un.* **39**, 702 (1958).
3. PHILIP, J. R., General method of exact solution of the concentration-dependent diffusion equation, *Aust. J. Phys.* **13**, 1 (1960).
4. KLUTE, A., A numerical method for solving the flow equation for water in unsaturated materials, *Soil Sci.* **73**, 105 (1952).
5. PHILIP, J. R., Numerical solution of equations of the diffusion type with diffusivity concentration-dependent, *Trans. Faraday Soc.* **51**, 885 (1955).
6. PHILIP, J. R., Energy dissipation during absorption and infiltration: I, *Soil Sci.* **89**, 132 (1960).
7. LAMB, H., *Hydrodynamics*, 6th ed. Cambridge University Press, Cambridge, 1952.
8. CHILDS, E. C., and COLLIS-GEORGE, N., The permeability of porous materials, *Proc. Roy. Soc.* Series A, **201**, 392 (1950).
9. MARSHALL, T. J., A relation between permeability and size distribution of pores, *J. Soil Sci.* **9**, 1 (1958).
10. RICHARDS, S. J., and WEEKS, L. V., Capillary conductivity values from moisture, yield and tension measurements on soil columns, *Soil Sci. Soc. Amer. Proc.* **17**, 206 (1953).
11. PECK, A. J., The diffusivity of water in a porous material, *Aust. J. Soil Res.* **2**, 1 (1964).
12. YOUNGS, E. G., An infiltration method of measuring the hydraulic conductivity of unsaturated porous materials, *Soil Sci.* **97**, 307 (1964).
13. QUIRK, J. P., and SCHOFIELD, R. K., The effect of electrolyte concentration on soil permeability, *J. Soil Sci.* **6**, 163 (1955).
14. BODMAN, G. B., and COLEMAN, E. A., Moisture and energy conditions during downward entry of water into soils, *Proc. Soil Sci. Soc. Amer.* **8**, 116 (1944).

15. YOUNGS, E. G., Moisture profiles during vertical infiltration, *Soil Sci.* **84,** 283 (1957).

16. PHILIP, J. R., An infiltration equation with physical significance, *Soil Sci.* **77,** 153 (1954).

17. PHILIP, J. R., The theory of infiltration: 1. The infiltration equation and its significance, *Soil Sci.* **83,** 345 (1957).

18. PHILIP, J. R., The theory of infiltration: 2. The profile at infinity, *Soil Sci.* **83,** 435 (1957).

19. PHILIP, J. R., The theory of infiltration: 3. Moisture profiles and relation to experiment, *Soil Sci.* **84,** 163 (1957).

20. PHILIP, J. R., The theory of infiltration: 4. Sorptivity and algebra i (infiltration equations, *Soil Sci.* **84,** 257 (1957).

21. PHILIP, J. R., The theory of infiltration: 5. The influence of the initial moisture content, *Soil Sci.* **84,** 329 (1957).

22. PHILIP, J. R., The theory of infiltration: 6. Effect of water depth over soil, *Soil Sci.* **85,** 278 (1958).

23. PHILIP, J. R., The theory of infiltration: 7. *Soil Sci.* **85,** 333 (1958).

24. NIELSON, D. R., KIRKHAM, D., and VAN WIJK, W. R., Diffusion equation calculations of field soil-water infiltration profiles, *Proc. Soil Sci. Amer.* **25,** 165 (1961).

25. PARR, J. F., and BERTRAND, A. R., Water infiltration into soils, *Advanc. Agron.* **12,** 311 (1960).

26. WATSON, K. K., A note on the field use of a theoretically derived infiltration equation, *J. Geophys. Res.* **64,** 1611 (1959).

27. YOUNGS, E. G., Redistribution of moisture in porous materials: 1, *Soil Sci.* **86,** 117 (1958).

28. YOUNGS, E. G., Redistribution of moisture in porous materials: 2, *Soil Sci.* **86,** 202 (1958).

29. RUSSELL, E. W., *Soil Conditions and Plant Growth,* 9th ed. Longmans, London, 1961.

30. CHILDS, E. C., The scientific aspects of field drainage, *Sci. Progr.* **174,** 208 (1956).

30a. LUTHIN, J. N. (ed.), *Drainage of Agricultural Lands.* Amer. Soc. Agron., Madison, Wisconsin, 1950.

31. CHILDS, E. C., The water table, equipotentials and streamlines in drained land, *Soil Sci.* **56,** 317 (1943). (The first of a series of publications bearing the same title.)

32. MARSHALL, T. J., The diffusion of gases through porous media, *J. Soil Sci.* **10,** 79 (1959).

33. CURRIE, J. A., Gaseous diffusion in porous media. Part 2—Dry granular materials, *Brit. J. Appl. Phys.* **11,** 318 (1960).

34. CURRIE, J. A., Gaseous diffusion in porous media. Part 3—Wet granular material, *Brit. J. Appl. Phys.* **12,** 275 (1961).

35. PHILIP, J. R., and DE VRIES, D. A., Moisture movement in porous materials under temperature gradients, *Trans. Amer. Geophys. Un.* **38,** 222 (1957).

CHAPTER 7

Some Experimental Aspects of Crop Water Use Studies in the Field

7.1. The Measurement and Calculation of Terms in the Water Conservation Equation

Although interest may centre particularly on one or more component, crop water use studies often involve the determination of all the components of the water conservation or "water balance' equation (1.11) discussed in section 1.3. This equation is repeated below with the additional term for water input by irrigation I included. Another change from eqn. (1.11) is that evaporation rate E will be expressed in $cm\,sec^{-1}$ (or $cm^3\,cm^{-2}\,sec^{-1}$) units, so that all the terms in the equation may be considered as the depth of equivalent rainfall or surface ponded water in centimetres. Thus the conservation equation is:

$$P + I - S = \Delta D + \Delta M + \int E\,dt + U \quad (cm), \qquad (7.1)$$

where the meaning of the remaining terms are defined below eqn. (1.11). This water balance equation partitions the water applied to a given volume of soil in a given time period. While the depth of this soil volume may be arbitrary, it is desirable that this depth should exceed rooting depth when estimating evaporation from vegetation by measuring or calculating all other terms in eqn. (7.1), otherwise the estimate of evaporation may not be accurate.

The determination of the components of the water balance equation (7.1) will now be considered in the order in which they appear. Inputs P and I, and run-off S tend to be discrete, whilst the output terms evaporation ($\int E\,dt$) and through drainage U are continuous in character, as is also the change in storage term ΔM.

178

INPUTS (P AND I) AND SURFACE RUN-OFF S

Precipitation P can be measured by collecting rainfall in a sharp-rimmed cylindrical container designed to prevent loss by raindrop splash. Particularly in windy situations the turbulent airflow round a *rain gauge* can influence the catch to some extent. Turbulence decreases as the soil's surface is approached, but a rain gauge with rim at soil level would be inaccurate due to splash collected from the surrounding surface. As a compromise between these two sources of error, the British Meteorological Office recommends all gauges be mounted with their rim 12 in. above an extensive area of short grass, where this is feasible.

The spatial variation in rainfall amount, typically rapid with convectional type storms, poses the question of what spatial replication of rain gauges is necessary to estimate the total quantity of rainfall received over an extended area with a given degree of accuracy (Pereira *et al.*[1]). Similar problems arise regarding irrigation input I. Even where irrigation is carefully regulated there is a tendency for non-uniformity of accession with both spray and furrow irrigation. With sprays this can be due to the complex pattern of distribution of water over the field, and with furrow irrigation cumulative infiltration tends to decrease with distance from the head ditch, and this is accentuated by long furrows and high infiltration rates (Philip and Farrell[2]). Thus even if the value of I for the entire irrigated area is known, it will not necessarily represent the net accession of irrigation water at any particular point. Hence, unless the problem is overcome by adequate replication, I is an uncertain if not entirely unknown quantity in eqn. (7.1).

Hudson[3] has described structures and techniques used for measuring run-off from 1/20 acre plots. Partly because of the large volume of water which may be involved, the direct measurement of S can be difficult.

Thus, whilst there are problems, none insuperable, in determining I, P and S separately for the particular soil volume to which eqn. (7.1) is to be applied, the *net input*, defined as $(P + I - S)$, can be more readily determined as follows.

When water supply is not limited evaporation is controlled very largely by meteorological factors, and actual evaporation can be considered to approach "potential evaporation". Potential evaporation is defined as the evaporation from an extensive area of uniform low actively growing crop, not short of water, completely shading the ground (Penman[4]). Equation (3.17) in section 3.2 may be used to calculate this quantity. Therefore E can be estimated for the period from just before water accession to a later time when above surface storage has disappeared (i.e. $\Delta D = 0$), and before any limitation on evaporation has developed. The terms U and ΔM can be calculated from measurements as will be described later in this section. Thus applying eqn. (7.1) over this short period, net input $(P + I - S)$ can be obtained by solution as the only unknown.

The warning should be given that situations exist where advection of energy upwind from the site may cause actual evaporation to be greater than the potential rate for reasons discussed in section 1.4 and by Penman, Angus and van Bavel[5]. On the other hand, particularly if the evaporation potential is high, stomatal control may reduce actual evaporation to below the potential before water stress is general in the root zone (Denmead and Shaw[6]).

Surface detention changes (ΔD)

As indicated above, the time period of application of eqn. (7.1) can usually be chosen so that surface detention is zero or negligible both at the beginning and end of the period. If so then $\Delta D = 0$.

CHANGES IN SOIL-WATER STORAGE (ΔM)

In general volumetric water content θ varies with depth z beneath the soil surface, Fig. 40 giving some simple examples. The total water in depth of equivalent rainfall stored from the soil surface to a profile depth d can be calculated from $\int_{o}^{d} \theta \, dz$ (cm). This is given, with due attention to scale, by the area enclosed between the axes of θ and z and the volumetric water-content profile to the depth of interest (Fig. 40). The determination of such profiles and thus of ΔM is discussed in section 7.2.

Evaporation $\int E \, dt$

Because of the experimental difficulty in accurately and directly measuring evaporation (or "evapotranspiration") from crops in the field, it is frequently sought by solution of the water conservation equation as the only unknown. However, there is at present no convenient *direct* method of measuring the through drainage U below root depth in field experiments when evaporation is also unknown. Thus there are in general *two* unknowns, U and $\int E \, dt$, so that unless one of these can be separately determined only $(U + \int E \, dt)$ can be obtained by this approach. Many estimates of evaporation have been too high (if U is positive), or too low (if U is negative, indicating upward movement of soil water), through neglecting U. Although there can be situations where the neglect of this term may lead to small errors over some restricted time interval, the evidence indicates (e.g. Wilcox[7]) that this is generally not the case. A method of calculating U from other measurements is discussed later in this section.

If a representative sample of the plant community is grown on a volume of soil isolated from its surroundings by a small air space so that it can be weighed, evaporation can be directly measured by the loss in weight. Although isolating the soil column by enclosing it in a container significantly alters the lower boundary conditions, provided soil depth is as great as the rooting depth, differences between evaporation from plants within the isolated soil column and outside it can be small. Such an installation is often called a *weighing lysimeter*, a *lysimeter* being an unweighed isolated soil column with provision for sampling percolate collected beneath the column. Slatyer and McIlroy[8] have discussed important factors affecting the design of such installations. Development in simplicity and accuracy in weighing lysimeters is still taking place, and examples of existing installations have been described by King et al.,[9] Pruitt and Angus,[10] McIlroy and Sumner,[11] van Bavel and Myers,[12] and Rose, Byrne and Begg,[13] for example. An accuracy in evaporation measurement equivalent to 0·002 cm rainfall can be achieved with some types of weighing lysimeter.

In section 2.3 the fluctuation theory basic to the "eddy-correlation" approach to evaporation measurement (Taylor and Dyer[14]) was described. Whilst a humidity sensing device with shorter response time would improve the accuracy of this equipment, the possibility of portable evaporation measuring equipment based on this approach is most attractive in agricultural research.

Potential evaporation, defined earlier in this section, provides a useful baseline against which to compare actual evaporation. The definition given by Penman (loc. cit.) to potential evaporation is often extended to crops other than short grass, the evaporation then depending to some extent on the crop in question (Mather[15]). Evaporation from a particular crop with water plentifully supplied can be simply measured over periods of a month or so by measuring rainfall and irrigation input and leachate extraction. Over a sufficiently long period ΔM is small compared with other terms in eqn. (7.1), thus yielding evaporation with at least useful accuracy. Slatyer and McIlroy (loc. cit.) have described such measurements using what they term a "potential evaporimeter".

Through drainage U

As shown by Rose and Stern,[16] and by Holmes,[17] this term can be obtained by integrating over time water flux values (v) calculated by an application of the theory given in section 6.1. The equations leading to eqn (6.6) describing liquid water movement show that water flux v is related to the gradient of the soil-water suction profile ($\partial h/\partial z$) by:

$$v = K\frac{\partial h}{\partial z} - K \quad (\text{cm sec}^{-1})$$

with z taken as positive *upwards*, or

$$v = K\frac{\partial h}{\partial z} + K \quad (\text{cm sec}^{-1}) \tag{7.2}$$

with z positive *downwards*. Unless the soil-water suction (h, cm) is quite low, the suction gradient is usually far greater than unity, so that the term K in eqn. (7.2) is often negligible in comparison with $K(\partial h/\partial z)$ under such circumstances. The through-drainage component U in eqn. (7.1) then follows from:

$$U = \int_0^T v \, dt, \tag{7.3}$$

where T is the period over which eqn. (7.1) is applied. Assuming a soil to possess a uniform profile the upper limit to the magnitude of v is set by saturated flow under the action of the gravitational potential gradient alone. The value of v is then equal to the hydraulic conductivity at saturation in the z-direction [eqn. (7.2) with $(\partial h/\partial z) = 0$].

From eqns. (7.2) and (7.3) it follows that to calculate U, the slope is required for the soil-water suction profile ($\partial h/\partial z$) at the lower boundary of the soil volume to which eqn. (7.1) is applied. Also the hydraulic conductivity K at that depth, and how it varies with water content, needs to be known. The determination of K is described in section 7.2. It follows from Fig. 33 that ($\partial h/\partial z$) could be obtained from a knowledge of both the water-content profile and the moisture characteristics at, above, and below the depth of the lower boundary at which U is required. Particularly at lower suctions (Fig. 33) hysteresis in the soils' moisture characteristics should be taken into account. However, the effects of hysteresis can only be accounted for approximately unless the change in water content has been large enough and in the same sense to erase the effects of the past history of such changes (Childs[18]), in which circumstances the unique outer envelope of moisture characteristics (Fig. 33b) is appropriate for the relationship between suction and water content.

As the water content approaches saturation it becomes increasingly difficult to achieve adequate accuracy in the suction gradient

$(\partial h/\partial z)$ for the calculation of water flux v if h is inferred from water content via the moisture characteristics (Rose and Stern, *loc. cit.*) This problem can be overcome by measuring h directly using tensiometers (Figs. 29 and 30). Soil can remain saturated over a limited range of suction. Soil-water suction in saturated soil must be directly measured since it cannot be inferred from water content.

The direct measurement of h at lower suctions using tensiometers also has the advantage that it includes the effect of overburden pressure B on soil-water suction, a topic discussed in section 5.7. It follows from eqns. (5.19), (5.20) and (5.21), that soil-water suction h is given approximately by:

$$h = (s - aB)/\rho g \quad \text{(cm)}, \tag{7.4}$$

where s (dyne cm^{-2}) is the unloaded suction, a a load partitioning coefficient, ρ the density of water (g cm^{-3}), and g the acceleration due to gravity (cm sec^{-2}). Overburden pressure B may be calculated from the profile of bulk density ρ_b and volumetric water content θ above the point where h is required. Taking z as zero at the soil surface and positive downward, the value of B at depth z (cm) is given by:

$$B = \int_0^z (\rho_b + \rho\theta)g \, dz \quad \text{(dyne cm}^{-2}\text{)}. \tag{7.5}$$

It is really unloaded suction s, and not *in situ* soil-water suction h, that is obtained on entering the moisture characteristics (Fig. 33) with water content measured in the field.

Thus, if soil-water suction *in situ* h is not directly measured, but obtained indirectly from water content, h should in general be calculated from unloaded suction s using eqns. (7.4) and (7.5). The determination of coefficient a in eqn. (7.4) will be described in section (7.2). The correction term in B [eqn. (7.4)] due to overburden pressure will become negligible compared to h when h becomes sufficiently high, but just how high depends on the value of a (i.e. on soil type and water content) and on the magnitude of B (i.e. on the depth below the soil surface, and the bulk density and water-content profiles above that depth [eqn. (7.5)]). Usually, the correction due to B will only be of importance at suctions sufficiently low that h can be measured with a tensiometer (i.e. up to about $0 \cdot 6$–$0 \cdot 8$ atmospheres—see section 5.7).

Conclusion

As mentioned before, the water conservation eqn. (7.1) is commonly used to obtain evaporation by solution as the only unknown. Particular difficulty has been associated with the determination of the through-drainage component U. However, it can be calculated as described above provided the variation down the profile of the following soil properties is determined: (1) Bulk density. A knowledge of this property is not required if volumetric water content is measured directly using the neutron moderation technique described in the next section, except in connection with calibrating this technique. For swelling soils the relationship between bulk density and water content must be considered if the bulk density of such soils is used, as will be discussed in section 7.2. (2) The relationship between hydraulic conductivity and water content, including the value at saturation. (3) The moisture characteristics of soil for both wetting and drying. (4) The relationship between the coefficient a and water content. Relationships (3) and (4) are not required if suction h is measured directly with tensiometers.

The determination of these soil properties, and also of volumetric water content, will now be described.

7.2. Determination of the Soil Properties required in Water Balance Studies

VOLUMETRIC WATER CONTENT

This can be determined by measuring gravimetric water content w (grams of water per gram of dry soil), then separately measuring bulk density as described later, and calculating θ using eqn. (5.7). The techniques and replication due to soil and thus water-content variability have been examined by Slatyer and McIlroy (*loc. cit.*). A useful sampling technique is to bulk soil samples from the same depth in replicate holes (in a billy-can with lid, for example), mix well, and then take a sub-sample for water content determination in a tin with tight fitting lid (tobacco tins with screw-action lid closure are suitable). Particularly if temperatures are high, exposure of soil to the air should be kept as brief as possible, and filled sample tins should be kept in the shade until weighed.

The electrical resistance of various porous media inserted in soil, of sufficiently small size to be at least in approximate suction equilibrium with the soil, has found considerable use as an indirect method of following water-content changes at a particular location in the soil. Bouyoucos,[19] Bourget *et al.*[20] and Far-brother,[21] for example, have described the construction and operation of such instruments. Possibly because of the necessity for individual calibration (or at least check calibration) on each unit, and because of the lower precision, this electrical resistance technique is in some applications being superseded by that of fast neutron scattering, now to be described.

Fast or energetic neutrons are produced or liberated during certain nuclear reactions. A commonly used fast neutron source consists of a mixture of radium and beryllium. The a-particles or helium nuclei resulting from the radioactive decay of one of the radium isotopes can react with beryllium nuclei, one of the products of this reaction being an energetic neutron. Because neutrons are particles bearing no electrical charge, neutrons of all energies have free access to atomic nuclei with which they can interact in two general ways. The interaction can result either in capture of the neutron by the nucleus with which it interacts, or in scattering, either with or without loss of energy by the scattered neutron. The energy loss in scattering increases as the mass of the nucleus approaches that of the incident neutron, hydrogen nuclei thus being the most effective in slowing down fast neutrons. When slowed down to speeds comparable with those of the thermal motion of surrounding nuclei, the neutrons then scatter or diffuse with little further energy loss, ultimately being captured however.

Fast neutrons emitted into soil will be slowed down principally by hydrogen nuclei, and thus by water in the soil. If organic matter content is high this may also be a significant contributor to neutron moderation. The flux of slow neutrons in the region of a fast neutron source in soil will thus be most strongly affected by the water content of the soil and the strength of the source (or the initial rate of fast neutron release). The neutron scatter technique of measuring soil water thus consists essentially of lowering a fast neutron source and a slow neutron detector into a hole made in the soil, and measuring the slow neutron flux scattered back to the detector. Detector and source are conveniently enclosed in the same container, which can be lowered down the hole with the electric cable required to supply power to the detector and collect the pulses from it which give a measure of the slow neutron flux. The average rate at which these pulses are produced can then be measured by counting the number of pulses produced over a measured period of time with the aid of an electronic counter, usually referred to as a "scaler" since it commonly counts pulses and displays the cumulative count in scales of ten. Alternatively, the rate of pulse production can be measured directly with a "ratemeter". Although the accuracy of ratemeter measurement is inherently less than with counting pulses over a fixed period of time, or timing a fixed number of pulses, the performance of good equipment may be adequate for many purposes.

Although usually only of secondary importance in soils, the flux of slow neutrons is also significantly affected by neutron capture. Chlorine, iron and boron, for example, are effective neutron absorbers, but as these elements are frequently only present in small amounts in soil, the slowing down of fast neutrons due to scattering from hydrogen nuclei is the factor normally dominating slow neutron flux in soils. When this is so, from the point of view of production of slow neutrons the source is effectively surrounded only by the water in the soil. Thus the relationship between slow neutron count rate and volumetric water content for a particular source and detector (line A, Fig. 43) is normally similar for different soils. For reasons mentioned above, high levels of neutron absorbers such as chlorine and boron can appreciably reduce the neutron flux and thus the slow neutron count rate. Since neutron absorption is proportional to neutron flux, the effect of neutron absorption is to reduce slow neutron count rate by a constant fraction. As demonstrated by Holmes and Jenkinson[22] this reduces the slope of the count rate versus water content relationship, as illustrated in line B, Fig. 43, a 10 per cent reduction

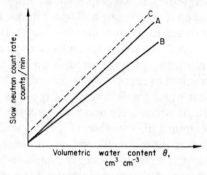

FIG. 43. Effect of increased neutron absorbers (curve B) and more bound water (curve C) on the relation (curve A) between slow neutron count rate and volumetric water content (on an oven-dry basis).

in slope in the calibration curve being caused by the addition of about 100 ppm of boron, or 7400 ppm of chlorine (or 0·012 g NaCl/g dry soil). If such elements are present in atypical abundance in soil the modified calibration curve appropriate to the particular soil should be employed. Holmes and Jenkinson (*loc. cit.*) indicate how such modification can be determined from the concentration and neutron absorption properties of elements. Changes in soil density have similar but smaller effects, an increase from 1·4 to 1·6 g/(cm³ dry soil) causing a 2·4 per cent decrease in slope of the calibration curve (Holmes and Jenkinson, *loc. cit.*).

As the clay content of a soil increases so does the amount of water which is so strongly bound to the clay lattice by electrical forces that it is not removed by oven drying at the normal temperature of 105°C. Since calibration curves are plotted against water content determined by oven-drying (Fig. 43), an increase in bound water will shift the calibration curve A upwards by a constant amount equal to the count rate of neutrons slowed down by the bound water (curve C, Fig. 43). Soils with higher amounts of bound water also tend to contain larger proportions of neutron absorbers, resulting in a calibration curve somewhere between C and B in Fig. 43, perhaps fortuitously being not very different from A. Thus soils of different types can have similar calibration curves, but this needs to be checked by a calibration procedure which will be briefly described after discussing the size of the spherical volume sampled by the neutrons.

Denote by R (cm) the effective maximum radius for neutron moderation measured from the source and detector. Since there must be approximately the same number of collisions between fast neutrons and hydrogen nuclei before the neutrons are slowed down, then $R^3\theta \doteqdot$ constant. Although the water in soil close to the neutron source has a greater effect on the counting rate than has soil water farther away, the size of soil volume which influences the counting rate of slow neutrons may be approximately estimated from:

$$R \sim 12\theta^{-1/3} \quad (cm). \tag{7.6}$$

If $\theta = 0.2$, then from eqn. (7.6), $R \sim 20$ cm. Thus readings should not be taken with source and detector closer to the soil surface than about 20 cm at this water content, otherwise they will be in error due to fast neutrons escaping from the soil surface. (However, it is the surface layer where water content can be most easily determined by gravimetric sampling.) Furthermore, rapid profile changes in water content tend to be smoothed out because the scattered neutron flux reflects a weighted average of water content over a certain volume. Holmes and Jenkinson (*loc. cit.*) give an example of such smoothing.

Calibration of a neutron moisture meter in a particular soil may be carried out either in the field or with soil brought into the laboratory. In field calibration a number of access holes are augered to the desired depth of investigation, and as in all measurements it is desirable to line the access hole with, for example, polythene or aluminium tubing closed at the lower end. Such tubing has no measurable influence on counting rate. Count rate is then measured at a range of depths and water contents, gravimetric water content being determined on samples retrieved

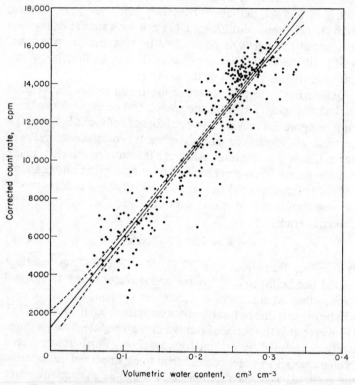

FIG. 44. Field neutron moisture meter calibration for Ord sandy loam, a soil in Western Australia.

by augering close to the access tube, and to the same depth as the source and detector. The bulk density profile must then be determined for the soil *in situ* as described later in this section, and used to convert gravimetric to volumetric water content. Figure 44 illustrates the type of results obtained using field calibration. The narrow confidence limits for the mean curve fitted to the data were obtained in the presence of the scatter indicated only by fairly extensive replication. For this reason it is probably more efficient to build up the calibration curve using soil brought to the laboratory, though considerable effort is required for accurate results in either method.

In laboratory calibration the soil is thoroughly mixed, so as to be as uniform in water content as possible. It is then packed into a cylindrical container or drum with dimensions not less than $2R_{max}$, where R_{max} is the expected maximum value of R in eqn. (7.6), corresponding to the minimum value of θ. The drum should be centrally fitted with a section of the same access tube liner which is used in the field, to receive the probe. After the count rate has been determined, a number of subsamples from soil near the probe should be taken for determination of water content. Over some range of water content soil samples of known volume can be extracted using equipment similar to that used in bulk density sampling (described later). Volumetric water content can then be calculated from gravimetric water content and bulk density determined for the *same* sample. This yields an improvement in accuracy over field calibration where gravimetric water content and bulk density are determined on *different* samples. Uniformity in water content can also be greater in laboratory calibration, leading to more accurately known water content.

It is often desirable to obtain a calibration at a water content so high that it is very difficult if not impossible to remove a soil sample of accurately known volume and the same water content as the soil in the container. For a saturated sample the following method has been used by the author: Using the cylindrical soil container as a mixing bowl gradually add weighed quantities of air-dry soil and water, mixing to a uniform saturated consistency, until the container is exactly full. Knowing the container volume, the volumetric water content can be directly calculated.

The aim is that the calibration should apply to the soil *in situ*, which in general will be at a bulk density different from that used in the laboratory calibration. As indicated above, the effect of such differences are calculable and small in themselves, but indirectly the effect may be greater, especially in soils with a relatively high proportion of bound water. Whilst an increase in bulk density decreases the slope of the calibration curve slightly, the greater amount of bound water associated with the higher bulk density often gives rise to a net upward shift in the calibration curve with increasing bulk density. This problem can be experimentally investigated by uniformly packing air-dry soil to two different bulk densities. A high bulk density can be achieved by reducing the size of soil aggregates before filling the container.

Using the information from such an experiment, together with the measured calibration and *in situ* bulk densities, the laboratory calibration data can be transformed into a calibration for field use.

BULK DENSITY

Most manufacturers of neutron moisture-measuring equipment supply a probe for the measurement of bulk density compatible with the same particle-counting equipment. A bulk density probe consists of a source and detector of γ-radiation which can be let down the same moisture probe access tube by means of the cable supplying power to and collecting electrical pulses from the detector.

The source chosen emits γ-radiation of such an energy that attenuation is predominantly due to elastic "Compton" scattering, which exhibits the feature of being independent of the chemical nature of the bombarded element up to an atomic number of 26, except for hydrogen. Within this range, which includes the most abundant elements present in soil, radiation absorption is directly proportional to the density of the absorbing material. The absorption of γ-radiation by hydrogen is twice as high as other common elements in soil, which would lead to errors in bulk density due to variation in water content. This error is commonly attenuated by a circuitry technique which discriminates against low-amplitude pulses coming from the detector, or by a judicious use of lead screening to reduce low-energy radiation reaching the detector.

Bulk density can also be determined by carefully cutting and extracting a soil sample of measured volume, oven drying, and weighing. Fox and Page-Hanify[23] have described sampling equipment suitable for use near the soil surface. Stace and Palm[24] detail equipment for sampling at depth, consisting essentially of steel tube with suitably shaped cutting edge. If a pit is dug with two opposite walls vertical, these sample tubes can be driven horizontally into one wall using a motor vehicle jack supported against the opposite wall. If the sample tube is oiled, soil extraction is much easier.

The bulk density decrease with increasing water content can be quite appreciable in expanding soils. The analysis of the two modes of swelling given by Fox[25] provides a theoretical basis for fitting curves to field data.

HYDRAULIC CONDUCTIVITY

Luthin,[26] and the American Society of Agricultural Engineers (Bouwer et al.[27]) have summarized the various methods currently available for measuring hydraulic conductivity *at saturation* in the field.

Hydraulic conductivity can be determined in the field over the entire range of water contents, whatever the nature of the soil profile, using a method described by Rose, Stern and Drummond,[28] and given below.

Consider a region of soil free of vegetation and covered at the surface, for example with a plastic sheet, to prevent any water flux across the soil surface. The water conservation equation (7.1) with all terms zero except ΔM and U, when applied to this volume of soil with the surface taken as depth zero and the lower boundary at depth z, is:

$$- U = \Delta M \quad \text{(cm)}$$

or

$$- \int_{t_1}^{t_2} v_z \, dt = \int_{t_1}^{t_2} \int_0^z \frac{\partial \theta}{\partial t} \, dz \, dt \quad \text{(cm)}, \qquad (7.7)$$

where t_1, t_2 = times of observation of consecutive water content
               profiles (sec),
       v_z = vertical flux of water at depth z (cm sec^{-1}),
        z = depth, positive downward, measured from the soil
               surface (cm), and
        θ = volumetric water content (cm^3 cm^{-3}).

Substituting $v_z = K_z(\partial h/\partial z) + K_z$ from eqn. (7.2) into eqn. (7.7)
gives:

$$\bar{K}_z = \left[-\int_{t_1}^{t_2}\int_0^z \frac{\partial \theta}{\partial t}\, dz\, dt \right] \Big/ \left[\overline{\frac{\partial h}{\partial z}} + 1 \right]_z T, \text{ (cm sec}^{-1}) \qquad (7.8)$$

where $\bar{K}_z$ is an average hydraulic conductivity over the time
interval T [$= (t_2 - t_1)$ sec] between measurements of successive
water-content profiles. The implicit assumption is that the time
taken in measuring a profile of water content is small in comparison with T. This assumption is easily satisfied in practice using
a neutron moisture probe. If gravimetric sampling were used to
determine water-content profiles the requirement of this assumption would be difficult to meet under near saturation conditions
in a permeable soil. (Gravimetric sampling also introduces the
problem of having to sample at different sites in successive profile
measurements.)

Graphically the term ΔM or

$$\int_{t_1}^{t_2}\int_0^z (\partial \theta/\partial t)\, dz\, dt$$

in eqn. (7.8) is the area enclosed between the soil surface and depth
z, by water-content profiles plotted for times t_1 and t_2. The mean
suction gradient $\overline{(\partial h/\partial z)}$ is the mean of the suction gradients at
depth z at the two times t_1 and t_2. The determination of suction
gradients was fully discussed in section 7.1, and all the remarks on
methods, accuracy, and correction for overburden pressure given
there apply also in this context.

The interval T between successive profile water measurements should not be too small, or the term $\int\limits_{t_1}^{t_2}\int\limits_{0}^{z} (\partial\theta/\partial t)\,dz\,dt$ may not be known with sufficient accuracy; neither should it be too large, or the use of the mean suction gradient $(\overline{\partial h/\partial z})$ may lead to errors. Thus T should be varied in accordance with the general rate of

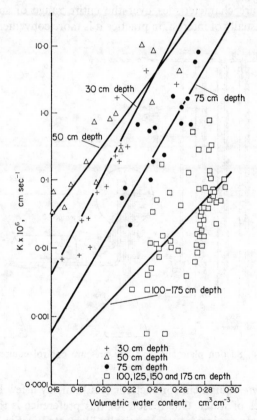

FIG. 45. Hydraulic conductivity characteristics of Ord sandy loam (after Rose, Stern and Drummond[28]).

change of water content. An example of a suitable sequence of values of T for the soil whose conductivity characteristics are shown in Fig. 45 is given by Rose, Stern and Drummond (*loc. cit.*), where T was extended as the period from irrigation increased.

SOIL MOISTURE CHARACTERISTICS

Whilst the pressure-membrane equipment described in section 5.9 and Fig. 35 is in principle suitable for the determination of soil-moisture characteristics over the entire range of soil-water suctions usually of interest, in practice it is more convenient to use

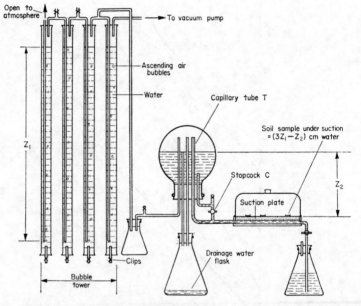

FIG. 46. Suction plate and ancillary pressure control equipment (not to scale).

suction plate equipment for suctions less than say a few hundred centimetres of water. One reason for this preference is that with suction plate equipment, schematically illustrated in Fig. 46, soil can be removed for water-content determination whilst the soil remains under suction, removing any doubt about error due to water redistribution between pressure release and sample removal, even though this is normally negligible. Another reason for preferring suction plate type of equipment at low suctions is the simple and accurate means of controlling and measuring suction using bubble towers (Fig. 46).

The suction plate itself may consist of fine-pored unglazed ceramic, sealed to a water chamber on its underside. The suction plate acts exactly as a tensiometer described in section 5.7, except that in a suction plate the applied suction controls the water content of the soil instead of the water content in the soil controlling the suction of the water in the tensiometer. The suction plate should normally be kept covered to prevent it drying out and air entering the system.

Figure 46 illustrates one possible arrangement for controlling the suction of the water in the suction plate, allowing exchange of water between the samples and the plate, and providing indication of when suction equilibrium has been achieved between the two. Evidence of the latter is provided if there is no movement of water in the capillary tube T with stopcock C closed. Suction of the water in the suction plate can be varied by connecting to different segments of the bubble tower chain.

COEFFICIENT a

Although there is some disagreement over the adequacy of the theoretical basis of the method, due to Coleman and Croney,[29] Black et al.[30] have shown experimentally that a can be approximated satisfactorily using the equation:

$$a = (\rho/\rho_b{}^2)(d\rho_b/dw)_{B=0} \tag{7.9}$$

where ρ = density of water,
ρ_b = soil bulk density, and
w = gravimetric water content.

The suffix $B = 0$ to the term $(d\rho_b/dw)$ indicates zero overburden pressure. This differential may be obtained experimentally by forming the wet soil into a simple geometrical shape and following the decrease in mass and size as the unloaded sample dries out. The size decrease may be followed optically. In general a increases with water content for an expandable soil in the manner illustrated in Fig. 47.

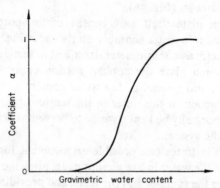

FIG. 47. The type of dependence of load partition coefficient a (see text) on soil-water content.

Bibliography

1. PEREIRA, H. C., McCULLOCH, J. S. G., DAGG, M., HOSEGOOD, P. H., and PRATT, M. A. C., Assessment of the main components of the hydrological cycle, *E. Afr. Agric. For. J.* **27**, 8 (1962).
2. PHILIP, J. R., and FARRELL, D. A., General solution of the infiltration—advance problem in irrigation hydraulics, *J. Geophys. Res.* **69**, 621 (1964).
3. HUDSON, N. W., The design of field experiments on soil erosion, *J. Agric. Engng. Res.* **2**, 56 (1957).
4. PENMAN, H. L., Evaporation: an introductory survey, *Neth. J. Agric. Sci.* **4**, 9 (1956).
5. PENMAN, H. L., ANGUS, D. E., and VAN BAVEL, C. H. M., Microclimatic factors in irrigation; Chapt. 27 in *Irrigation of Agricultural Lands*. Amer. Soc. Agron. (in press).
6. DENMEAD, O. T., and SHAW, R. H., Availability of soil water to plants as affected by soil moisture content and meteorological conditions, *Agron. J.* **45**, 385 (1962).
7. WILCOX, J. C., Rate of soil drainage following an irrigation. II Effects on determination of rate of consumption use, *Canadian J. Soil Sci.* **40**, 15 (1960).
8. SLATYER, R. O., and McILROY, I. C., *Practical Microclimatology*. Published by Commonwealth Scientific and Industrial Research Organization, Australia, for UNESCO, 1961.

9. KING, K. M., TANNER, B., and SUOMI, V. E., A floating lysimeter and its evaporation recorder, *Trans. Am. Geophys. Un.* **37**, 738 (1956).
10. PRUITT, W. O., and ANGUS, D. E., Large weighing lysimeter for measuring evapotranspiration, *Transactions of the ASAE*, **3**, 13 (1960).
11. McILROY, I. C., and SUMNER, J., A sensitive high capacity balance for continuous automatic weighing in the field, *J.Agric. Eng. Res.* **6**, 252 (1961).
12. VAN BAVEL, C. H. M., and MYERS, L. E., An automatic weighing lysimeter, *Agric. Engineering*, **43**, 580 (1962).
13. ROSE, C. W., BYRNE, G. F., and BEGG, J. E., An accurate hydraulic lysimeter with remote weight recording, C.S.I.R.O. Division Land Res. & Regional Survey; Technical Paper No. *27* (1966).
14. TAYLOR, R. J., and DYER, A. J., An instrument for measuring evaporation from natural surfaces, *Nature, Lond.* **181**, 408 (1958).
15. MATHER, J. R., *The Measurement of Potential Evapotranspiration*, Johns Hopkins University, Laboratory of Climatology, 1954.
16. ROSE, C. W., and STERN, W. R., The through drainage component of the water balance equation, *Aust. J. Soil. Res. 3*, 1 (1965).
17. HOLMES, J. W., Water balance and the water table in deep sandy soils of the upper south east, South Australia, *Aust. J. Agric. Res.* **11**, 970 (1960).
18. CHILDS, E. C., The ultimate moisture profile during infiltration into a uniform soil, *Soil Sci.* **97**, 173 (1964).
19. BOUYOUCOS, G. J., Improved soil moisture meter, *Agr. Eng.* **37**, 261 (1956).
20. BOURGET, S. J., ELRICK, D. E., and TANNER, C. B., Electrical resistance units for moisture measurements: their moisture hysteresis, uniformity and sensitivity, *Soil Sci.* **86**, 298 (1958).
21. FARBROTHER, H. G., On an electrical resistance technique for the study of soil moisture problems in the field, *Emp. Cotton Grow. Rev.* **34**, 71 (1957).
22. HOLMES, J. W., and JENKINSON, A. F., Techniques for using the neutron moisture meter, *J. Agric. Eng. Res.* **4**, 100 (1959).
23. FOX, W. E., and PAGE-HANIFY, D. S., A method of determining bulk density of soil, *Soil Sci.* **88**, 168 (1959).
24. STACE, H. T. C., and PALM, A. W., A thin walled tube for core sampling of soils, *Aust. J. Exp. Agric. and Animal Husb.* **2**, 238 (1962).
25. FOX, W. E., A study of bulk density and water in a swelling soil, *Soil Sci.* **98**, 307 (1964).
26. LUTHIN, J. N. (ed.), *Drainage of Agricultural Lands*. Amer. Soc. Agron., Madison, Wisconsin, 1957.
27. BOUWER, H. *et al.*, Measuring saturated hydraulic conductivity of soils, *Amer. Soc. Agric. Eng.*, Special Publication SP–SW–0262 (1962).
28. ROSE, C. W., STERN, W. R., and DRUMMOND, J. E., Determination of hydraulic conductivity as a function of depth and water content, for soil *in situ*, *Aust. J. Soil Res. 3*, 1 (1965).
29. COLEMAN, J. D., and CRONEY, D., The estimation of the vertical moisture distribution with depth in unsaturated cohesive soils, Note RN/1709, Road Research Laboratory, D.S.I.R., England, Unpublished mimeo, 14 pp. (1952).
30. BLACK, W. P. M., CRONEY, D. and JACOBS, J. C., Field studies of the movement of soil moisture. Tech. Pap. Rd. Res. Bd. No. 41. H.M. Stationery Office, London, 1958.

CHAPTER 8

A Physical Introduction to Plant–Water Relationships

8.1. Water Transport through the Soil–Plant–Atmosphere System

The extremely important questions of the relation between the uptake of water and plant nutrients and the role of water in plant growth processes will be neglected here and attention focused on the transpiration stream. Although water transport through plants can be under strong metabolic control in some circumstances (Kramer,[1] Mees and Weatherley[2]) the process appears to be a predominantly passive one (Slatyer[3]). The concept of water moving in response to gradients in its total potential Ψ, discussed in Chapters 5 and 6 in the setting of water and soil, is equally relevant to the passive transport of water through the entire soil–plant–atmosphere system (Van den Honert[4]).

Physically, the transpiration stream may be regarded as a water flux from a source of finite capacity, the soil water, down a gradient in total water potential, to a sink of effectively infinite capacity, the atmosphere. With roots in moist soil this gradient may fall to almost zero at the end of a night. But with the dawning of another day, water evaporated from the plant cells into the air spaces of the leaf diffuses out through the now opening stomata to the surrounding air (Fig. 48). The local decrease in Ψ accompanying this water loss causes upward flow through the xylem tissues of the stem, and subsequently through the roots. Thus absorption of water through the roots unavoidably lags behind the transpirational loss, and the excess of total transpiration over total absorption at any time gives the reduction in water stored in the plant. Decreasing water storage is reflected by lower turgor pressure within the plant cells and *vice versa*, and such turgor pressure fluctuations have a characteristic diurnal pattern under normal conditions.

The resistance r of any component to the flux q of water through the soil or plant may be defined by:

$$q = -\frac{\Delta \Psi}{r}. \tag{8.1}$$

Because of the obvious analogy between eqn. (8.1) and Ohm's law, it is convenient to represent the various soil and plant resistances

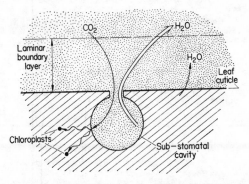

Fig. 48. Schematically illustrating the diffusion of water vapour through a stomate and the leaf cuticle away from the leaf. Inward diffusion of carbon dioxide for photosynthesis is also shown.

by a series arrangement of analogous electrical resistors (Fig. 49). For a given flux q it follows from eqn. (8.1) that the decrease in Ψ across any component, shown on the right of Fig. 49, is proportional to the resistance of that component. If this part of the figure were to scale, only the decreases in potential across the leaf and its boundary layer might be noticeable. The resistance of each component is indicated very roughly by the length of the component's symbolic representation.

The resistance r_{se} of the soil to the transpiration flux varies both because roots can extend to regions where Ψ is higher, and because of the dynamics of water absorption from soil by a root. The relative importance of these two processes needs investigation, and the complex nature of the latter process has given rise to a voluminous and often apparently conflicting literature on the "availability of soil moisture". Philip[5] and Gardner[6] have given a theoretical understanding of this problem, and Denmead and

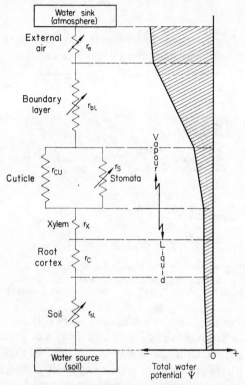

FIG. 49. (Left) Illustrating the various soil, plant and atmosphere resistances to the transpiration stream. Whilst no resistance is fixed, those shown with an arrow are particularly variable. (Right) Showing the decreasing total potential of water as it moves from soil through a plant to the atmosphere (not to scale).

Shaw[7] have experimentally demonstrated how the *potential evaporation rate* (or the maximum possible evaporation rate for the particular situation and type of surface if water were nonlimiting) and soil-water suction interact to control transpiration rate. As is illustrated in Fig. 50, the ratio of transpiration to potential evaporation rate declined as soil-water suction increased, but this decline commenced at considerably lower suctions as the

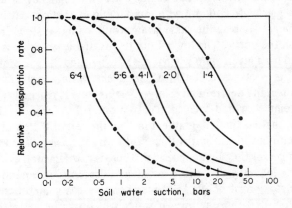

FIG. 50. Transpiration rate as a fraction of that with the soil at field capacity, plotted against soil-water suction. The curves represent this relation for days on which the transpiration rates at field capacity had the values shown by each curve in units of mm (24 hr)$^{-1}$ (after Denmead and Shaw[7]).

rate of potential evaporation increased. Since a drop in this ratio reflects a fall in turgor pressure and stomatal contraction, these data illustrate the dynamic character of plant–water relations referred to above.

Within the soil the problem involves an interaction between hydraulic conductivity K and hydraulic potential gradient $\partial\Phi/\partial s$ which may be considered using Darcy's equation (6.1). Consider the *slow* uptake of water by a root from soil of initially uniform water content. Since v (eqn. (6.1)) is small, $\partial\Phi/\partial s$ need also only be small. Thus K (a function of Φ) will be fairly constant with distance from the root at all stages of water removal, so that water will be withdrawn from a large soil volume, ensuring a greater and more constant supply of water to the root than in the case of rapid extraction now to be considered. Rapid absorption requires a large gradient $\partial\Phi/\partial s$, and so a large decrease in Φ and thus K close to the root. Absorption is therefore abruptly halted, leaving untapped water in the soil not far distant from the root. Lemon[8] has drawn an apt analogy with the tendency of a bent drinking straw to collapse with rapid liquid extraction, cutting off the supply, whereas the same straw would enable a slow drinker to finish the bottle. The lower limit to "available water" has been reached for a given plant when it has so exhausted the soil moisture round its roots as to be in a *permanent wilting* condition. As experimentally determined (Slatyer[9]) this state is reached when a plant remains wilted even though in a non-transpiring condition. Ideally, there is then no gradient in Ψ in the soil or the plant. The water content in the soil is then said to be the *permanent wilting percentage* for the plant concerned. There is experimental evidence (Slatyer, *loc. cit.*) that this percentage does not correspond to a unique suction of about 15 atmospheres as has sometimes been suggested, and the quantity of water which can be extracted at suctions greater than this figure appears to vary considerably with plant species, root distribution, and environmental and soil characteristics.

The absorption of water from the soil solution by plant roots is normally greatest in the region of *root hairs*, situated in a limited zone behind root tips. Thus, as the root system expands new root hairs are developed and older root hairs die and decay. Root hairs may be regarded as a special cell growing out from the root surface. In common with other cells the root hair has a *cell wall* consisting chiefly of cellulose lined on its inside with *cytoplasm*. This is a "living" part of the cell possessing amongst many other properties that of differential permeability (or conductivity) to water and solutes, normally being greater for water. In mature cells there is a proportionately large central region inside this cytoplasmic lining, called the *vacuole*, which is filled with *cell sap*. This is a watery fluid of variable molecular and ionic content, consisting of a solution or dispersion in water of mineral salts, sugars, organic acids, etc. In general there will be a difference in osmotic potential (section 5.12) between water in the soil solution bathing the root hair and water in the vacuolar sap of the hair, and it is clear that the osmotic as well as hydraulic component of the total potential of water must be considered, and this is taken up in section 8.2.

Water entering the root hair is transmitted down a gradient in its total potential (provided the process is passive) through the root *cortex* and cells of the *stele* to the *xylem vessels*, which are the principal channels of water movement from root to leaves. It is convenient to regard the resistance offered by the plant to liquid water movement through it as having the two components of root cortex resistance r_c, and xylem resistance r_x (Fig. 49). Except in quite wet soils, the resistance r_{sl} of the soil to the transpiration flux usually exceeds r_c and the normally quite small r_x (Gardner and Ehlig[10]).

Resistance to the transpiration stream when the flux is in the vapour phase was considered in 1900 by Brown and Escombe.[11] They also sought to understand why a photosynthesizing leaf in still air could absorb carbon dioxide at almost the same rate as a strong carbon dioxide absorbing solution with the same surface area as the leaf. Because stomata occupy only about 1 per cent, or at most some 3 per cent of the surface area of the leaf, so great an absorption of carbon dioxide by leaves at first sight presented something of a puzzle. The main key to this puzzle was suggested by Maskell,[12] and quantitatively expressed by Penman and Schofield[13] half a century after Brown and Escombe's work. This key lies in the recognition that a thin boundary layer, described in section 2.2 and which sheathes all solid surfaces including plant leaves, offers a considerable resistance to gas transfer across it. Air flow over at least some of the boundary layer will be laminar (section 2.2), so that gas or vapour transfer between leaf surface and the external turbulent air across this layer can be only by molecular diffusion. Figure 51 will be used in discussing the main features of gas exchange between plant leaves and their environment, and it shows an open stomate represented in model form by a cylinder of length l and radius a as in Fig. 48. The complex geometry of actual stomata (Heath[14]) makes this model approximate, but it is adequate for the present purposes.

Applying Fig. 51 to the transpiration stream, water evaporates into the sub-stomatal cavity and out through the stomate. The

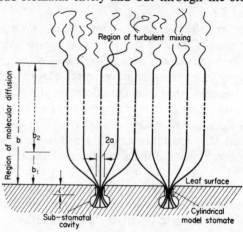

FIG. 51. Model of a leaf surface in cross-section, schematically illustrating diffusion of water vapour through stomata and laminar boundary layer out into the region of turbulent mixing.

separation of stomata is very variable, but is generally of the order of ten stomatal diameters. Hence the diffusion stream will expand, the lines shown in Fig. 51 indicating the mean flow direction. Merging of the diffusion streams from each stomate thus takes place at a short distance b_1 from the leaf surface, diffusion then being effectively one-dimensional for distance b_2, where $b_2 \gg b_1$ under agricultural conditions.

Diffusion is described by the Fick's law equation (6.14), where the flux density:

$$q_v = \frac{\varphi_v}{A} = - D_o \frac{\partial \rho_v}{\partial z} \quad (\text{g cm} \quad \text{sec}^{-1}),$$

with
φ_v = vapour flux (g sec^{-1}) across area A (cm^2)

D_o = diffusion coefficient of water vapour in air (cm^2 sec^{-1}), and

ρ_v = vapour density (g cm^{-3}).

Assuming one-dimensional diffusion in the direction z, and integrating this equation between z_1 and z_2 gives:

$$\varphi = \frac{D_o(\rho_{v_1} - \rho_{v_2})}{[(z_2 - z_1)/A]} \quad (\text{g sec}^{-1}), \tag{8.2}$$

where ρ_{v_1} is the vapour density at $z = z_1$.

The denominator $[(z_2 - z_1)/A]$ (cm^{-1}) of eqn. (8.2) may be referred to as a resistance to diffusion r, i.e.

$$r = (z_2 - z_1)/A \quad (\text{cm}^{-1}). \tag{8.3}$$

Although the vapour flux in transpiration (Fig. 51) is not completely one-dimensional, eqn. (8.2) will be applied to this process. From the above the diffusive resistance of a single stomate would be $(l/\pi a^2)$. However, in a situation such as Fig. 51, diffusing gas experiences an increased resistance over purely one-dimensional flow in having to diffuse up to a circular aperture, through a cylinder, and then away again. Because of this, the flow through a stomate can be correctly calculated if the effective length of the cylindrical stomate is taken to be greater than l by an amount $2x$, where more detailed considerations originally due to Stefan indicate that x should be $\pi a/4$ (Penman and Schofield, loc. cit.). The diffusion resistance due to stomata of leaf area A containing n stomata per unit area, and hence total number nA, may be obtained by regarding the resistance of each to be in "parallel" in the electrical sense of alternate pathways. It is thus:

$$\frac{(l + 2x)}{(nA\pi a^2)} = \frac{1}{An\pi a^2}(l + \pi a/2) \quad (\text{cm}^{-1}). \tag{8.4}$$

Suppose the model stomata in Fig. 51 represent cylindrical holes in a thin plate. Suppose further that the total area of holes per unit area of plate $(n\pi a^2)$ is kept constant, but radius a is decreased (so that n must increase). It follows from the right-hand side of eqn. (8.4) that the resistance of the perforated plate will also decrease quite appreciably, since a and l may be comparable in an open stomate. Browne and Escombe also obtained experimental evidence to support this deduction. This result, that the diffusion rate through a large number of narrow pores is greater than through a few wide pores, total pore area and other factors being the same in both cases, is modified to some extent by the diffusive resistance of the leaves' boundary layer which must also be traversed by diffusing molecules.

It need scarcely be added that Figs. 48 and 51 are highly schematic, and the rather prominent sub-stomatal cavity shown in these figures is often almost completely absent. Especially when partially closed, the stomatal pore is usually more slit-like than cylindrical, and very detailed examination of stomatal architecture is possible using the electron microscope. This has shown, for example, that in some plants the open length of the stomatal slit increases as the pore opens. Whilst the quantitative prediction of stomatal resistance from such anatomical information is possible (Jarvis, Rose and Begg, unpublished data), the cylindrical model is adequate for present purposes.

Let us now consider the total diffusive resistance experienced by the transpiration stream in the vapour phase until it reaches the turbulent air outside the laminar boundary layer of the leaf, where the resistance is no longer diffusive in character, and is much lower. From eqn. (8.3) the diffusive resistance of this laminar layer of depth b is given by (b/A) (cm^{-1}) for area A of leaf. Since this boundary layer resistance is in "series" with the leaf resistance given by eqn. (8.4), the total diffusive resistance for transpiration is:

$$(b/A) + (l + \pi a/2)/(An\pi a^2). \tag{8.5}$$

Substituting from eqn. (8.5) into eqn. (8.2), the diffusive flux per unit area q_v is given by:

$$
\begin{aligned}
q_v &= \frac{\varphi}{A} \\
&= D_o(\rho_{v_1} - \rho_{v_2})/[b + (l + \pi a/2)/(n\pi a^2)] \\
&= D_o(\rho_{v_1} - \rho_{v_2})/[r_{bl} + r_s] \quad (\mathrm{g\,cm^{-2}\,sec^{-1}}) \tag{8.6}
\end{aligned}
$$

where r_{bl} (cm) represents the boundary layer resistance, and r_s (cm) the stomatal resistance to diffusion of water vapour per unit leaf area. Diffusive resistances will hereafter be expressed per unit area, a change from the definition given in eqn. (8.3).

So far it has been tacitly assumed that transpiration is restricted solely to stomata and, unless stomata are closed, r_s is much smaller than the resistance of the leaf cuticle (Milthorpe[15]), which may be denoted r_{cu}. The parallel arrangement of resistances r_s and r_{cu} indicate that stomate and cuticle are alternate paths to the transpiration flux. What may be called leaf resistance r_l is therefore given by:

$$\frac{1}{r_l} = \frac{1}{r_s} + \frac{1}{r_{cu}}. \tag{8.7}$$

Total diffusive resistance (based on unit area) for the transpiration stream is thus:

$$r = r_{bl} + r_l. \quad \text{(cm)} \tag{8.8}$$

Whilst r_l is independent of wind speed, r_{bl} is not. The thickness of the leaves' laminar boundary layer, which may be taken as b in Fig. 51, and therefore the resistance r_{bl} of this layer to diffusion across it both decrease as the speed of bulk air over the leaf rises. However, at the relatively low wind speeds typical within plant communities, the resistance r_{bl} of the boundary layer usually exceeds leaf resistance r_l with open stomata (Slatyer and Bierhuizen,[16] who incorporate D_o in their definition of diffusive resistances). To illustrate the relative magnitude of stomatal and boundary layer resistances consider transpiration from a leaf of *Helianthus annus* in still air, as used by Brown and Escombe. From their data, with stomata open, $\pi a^2 = 9 \cdot 1 \times 10^{-7}$ cm^2 so that $a = 5 \cdot 5 \times 10^{-4}$ cm, $n = 33,000$ per cm^2, and $l = 1 \cdot 1 \times 10^{-3}$ cm. Thus from eqn. (8.6):

$$r_s = (l + \pi a/2)/(n\pi a^2) \quad \text{(cm)} \tag{8.9}$$
$$= 0 \cdot 065 \text{ cm in this example.}$$

Also from eqn. (8.6):

$$r_{bl} = b. \quad \text{(cm)} \tag{8.10}$$

Now under the most still air conditions obtainable indoors, and for surfaces about 10 cm in extent, b is approximately $1 \cdot 5$ cm (Penman and Schofield, *loc. cit.*). In still air r_{bl} is therefore more than an order of magnitude greater than r_s, and r_{bl} will still be

H

equal to or greater than r_s under some field conditions. Stomatal control of transpiration is thus effective in still air only if the stomata almost are closed, but in windy conditions favouring higher rates of evaporation open stomatal resistance may exceed that of the leaf boundary layer.

Direct and indirect methods of measuring stomatal aperture have been discussed by Heath (*loc. cit.*). A convenient type of instrument for obtaining a measure of stomatal aperture in the field is the viscous air-flow *porometer*. Such instruments measure the viscous resistance to air flow through a leaf, and Slatyer, Bierhuizen and Rose[17] have described a recent example. The output of the equipment described by van Bavel, Nakayama and Ehrler[18] is more directly related to stomatal diffusive resistance r_s.

The final resistance experienced by the transpiration stream is that of the external atmosphere (Fig. 49). As already mentioned, transport of water vapour in this stage is predominantly due to turbulent transfer described in section 2.3. This resistance was first considered in conjunction with plant diffusive resistances by Penman and Schofield (*loc. cit.*), and considered further by Monteith.[19]

As is illustrated in the plot of Ψ shown on the right-hand side of Fig. 49, most of the decrease in total potential of water composing the transpiration stream takes place when this water is in the vapour phase and passing from the *mesophyll cells* around the sub-stomatal cavity through the stomata to the outside air (Fig. 51). From eqn. (8.1) this implies that the greatest resistance to the transpiration stream takes place in this same region and with water in the same phase. Water in the liquid phase affects transpiration mainly indirectly, through the effect of loss of turgidity on stomatal movement. If the major control of transpiration were at a site in the plant earlier in the transpiration stream than the stomata, water in tissue beyond such a site would often be at such a low potential that the tissue could not survive (van den Honert, *loc. cit.*).

The disadvantages of employing divergent terminologies in different disciplines for the potentials and other concepts necessary to an understanding of plant–water relations become particularly obvious when the soil, the plant, and the atmosphere are considered as series components of a single system for the transpiration stream. Taylor and Slatyer[20, 21] have made unifying proposals based on thermodynamic terminology, and some of the difficulties associated with partitioning the total potential of soil water have been discussed by Bolt and Frissell.[22] The definition of potentials by the I.S.S.S.[23] discussed in detail in Chapter 5 is an example of this unified approach, and as stated in section 5.3 the total potential of water Ψ may be divided into the components defined in that chapter, whether the water is present in soil or a plant.

8.2. Water and Solute Transport across Cell Membranes

Suppose a container full of any liquid solvent (such as water) has introduced into it a membrane through which the solvent can pass, so that the solvent is divided into two regions. Suppose further that the total potential of either or both regions of solvent is altered by the addition of a solute to which the membrane is impermeable, so that a total potential difference is produced between the two regions. Then it follows from the general consideration explained in section 5.3 that a flux of solvent will be set up, in direction from the region of greater total potential to that of less, and this flux will tend to continue until the total potential difference is zero. This phenomenon is referred to as *osmosis*. Identical principles will control the movement of solutes, provided the effect of electrical potential gradients across the membrane are considered if the solute is ionic, and provided the solute transfer is *passive* or *non-active*. (In *active* transport, energy released by chemical changes can transfer molecules against the gradient in their total potential if the molecules are uncharged, or against the gradient in *electrochemical potential* if they are ionic (Dainty[24]).) Nevertheless, the term osmosis is usually, though not always, restricted to solvent transfer (Curtis and Clark[25]).

From Chapter 5 it follows that differences $\Delta \Psi$ in the total potential may be resolved into:

$$\Delta \Psi = \Delta \Phi + \Delta O, \qquad (8.11)$$

where Φ refers to hydraulic and O to osmotic potentials. Consider the simple osmometer of Fig. 52, in which the solvent is taken as

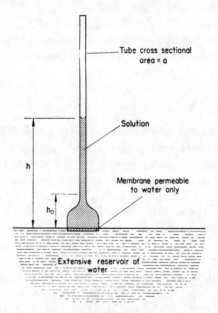

FIG. 52. A simple osmometer.

water. The difference in total potential on a unit volume basis, $\Delta \Psi_{vol}$, between water on the solution and water-reservoir side of the semipermeable membrane is the work required to transfer unit volume of water across the membrane from the reservoir to the solution side. From eqn. (5.10) and Fig. 52 it is given by:

$$\Delta \Psi_{vol} = \rho g h - \pi, \qquad (8.12)$$

where π is the osmotic suction of the solution, defined in section 5.12, and ρ the solution density, assumed constant over height h.

Thus it is seen that the effect of the addition of solutes is to lower the total potential of water in the solution. Initially, $\Delta\Psi_{vol}$ is negative, and water will flow from the region of higher total potential (the water reservoir) into the solution, thus increasing h and so $\Delta\Psi_{vol}$ (eqn. (8.12)). The inflow of water will continue until equilibrium is achieved when $\Delta\Psi_{vol}$ is zero, and ρgh is equal to π.

The achievement of equilibrium in this way results in some decrease in π due to dilution of the solution by the incoming water. Equilibrium could be achieved with no water influx, by the application of a pressure greater than atmospheric to the solution. The pressure at the solution side of the semipermeable membrane necessary to prevent any net flux of water across the membrane is called the *osmotic pressure* of the solution, and this is clearly identical with the osmotic suction π since it is immaterial whether equilibrium is achieved by increasing the hydraulic pressure on the solution side, or decreasing it on the water-reservoir side of the membrane.

From eqn. (5.8), the force per unit volume and thus the flux q of water will be proportional to $(-\Delta\Psi_{vol})$. Thus from eqn. (8.12),

$$q = K(\pi - \rho gh),\tag{8.13}$$

where K is a conductivity coefficient. Now solution volume:

$$V = V_0 + ah\tag{8.14}$$

where V_0 is the volume when $t = 0$, and $h = h_0$ (see Fig. 52). If A is the cross-sectional area of the membrane,

$$q = \frac{1}{A}\frac{dV}{dt},$$

which from eqn. (8.14)

$$= \frac{a}{A}\frac{dh}{dt}.\tag{8.15}$$

From eqns. (8.13) and (8.15), the equation describing the dynamics of the osmotic process is:

$$\frac{dh}{dt} = \frac{KA}{a}(\pi - \rho gh).\tag{8.16}$$

If the solution is sufficiently dilute for van't Hoff's law to apply:

$$\pi V = \frac{m}{M} R_u T, \qquad (8.17)$$

which is a constant here if absolute temperature T is constant, since m is the mass of solute of molecular weight M. Assuming eqn. (8.17) to be applicable, and making a further assumption concerning ρ, it is possible to obtain a solution of eqn. (8.16) expressing h as a function of time t.

The osmotic characteristics of plant cells will now be briefly considered. The main features of a typical unspecialized plant cell necessary for the present discussion were described in the previous section. Figure 53a illustrates a very young cell before vacuole formation, and Fig. 53b a mature cell, a large proportion of which

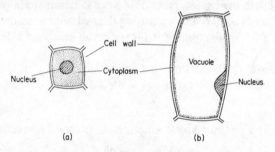

FIG. 53. (a) A very young plant cell mostly filled with cytoplasm, and (b) a mature cell with large central vacuole.

consists of the vacuole filled with the cell sap. The following discussion assumes a vacuole has been formed in the cell.

The cytoplasm contains various particulate bodies and is bounded by plasmatic membranes on both its cell wall and vacuole sides. It is these plasmatic membranes, and those at the boundaries of particular bodies which regulate the transport of cell substances (Dainty[26]). The cell wall (Fig. 53) is somewhat elastic, so that the volume of fluid within the cell can vary to some extent. The entry of water into the cell from outside it requires such an increase in cell volume, and this is accompanied by an increase in hydraulic pressure in the vacuole. The pressure increase depends both on the volume increase relative to that of the cell, on the elastic properties of the cell wall, and probably to some extent on restraint provided by surrounding cells. Particularly for a cell with relatively thin wall and large central vacuole, the hydraulic pressure gives the cell a mechanical rigidity it would otherwise lack, just as with an inflated balloon. A cell in which the pressure of the vacuolar fluid exceeds that of the atmosphere is said to be *turgid*, and the pressure above atmospheric is called the turgor pressure (T.P.). The wilting of a plant indicates that this pressure has fallen practically to zero in many cells. There is some evidence that under severe water stress T.P. may be negative (Slatyer[3]), thus having the character of a matric suction (section 5.6). Presumably in this state the cell would be *plasmolysed*. This term is used to describe the state of a cell in which turgor has been lost, and sufficient water withdrawn from the vacuole for the *protoplast* (cytoplasm and nucleus) to be at least partially contracted from the cell wall.

The ability to maintain a positive turgor pressure under suitable conditions is a characteristic of cell membranes in living tissue. To maintain turgor these membranes must be more permeable to water than to solutes. Understanding of the permeability of cell membranes is still very incomplete, but it is probably true to say that the transport of water across such membranes is predominantly passive, and the transport of solutes predominantly active. Assuming passive transport of water, and neglecting any interaction between water and solute fluxes (see p. 216), eqn. (8.12) can also be applied to water transport between the vacuole and the external solution bathing the cells. In this application the hydraulic pressure $\rho g h$ is replaced by the turgor pressure, and π is now regarded as the osmotic pressure (or suction) of water within the cell minus that of the solution outside. Water outflow from the cell will take place if $\Delta \Psi_{vol}$ is positive, i.e. if T.P. $> \pi$. Inflow of water will take place if T.P. $< \pi$, and equilibrium will correspond to T.P. $= \pi$. Since the membranes of plant cells are obviously not completely impermeable to solutes (even if their permeability to solutes is lower than to water), the statement that $\Delta\Psi$ controls water flux across the cell membranes may have to be modified because of interaction between solute and water fluxes (Dainty,[26] and Kedem and Katchalsky[27]). The flux of either water or solute in different situations may either aid or oppose the flux of the other component.

A considerable variety of terms have been used to denote the difference (π – T.P.), some common examples being *diffusion pressure deficit* (D.P.D.), suction force, and turgor deficit, such variety underlining the great desirability of a widely agreed terminology.

Yet another very important difference between plant cell behaviour and our previous assumptions is that *active transport*, in the sense of chemical coupling between fluxes and metabolism certainly occurs with solute transfer, although evidence for active water uptake by cells is not so well assured (Slatyer[3]).

Assuming *passive transport*, no interaction between solvent and solute fluxes, and a simple elastic model for cell-wall behaviour under turgor pressure, Philip[29] investigated the dynamics of water and solute fluxes in a cell-type system.

8.3. Conclusion

Although the discussion in section 8.1 has been limited to the transport of water, the discussion on water-vapour flux has obvious implications for the oppositely directed flux of carbon dioxide involved in assimilation. Both fluxes have to overcome the external air resistance r_e (Fig. 49), and, when allowance is made for the lower molecular diffusion coefficient of carbon dioxide compared with water vapour, both boundary layer and stomatal resistances r_{bl} and r_s also have to be crossed. However, as is indicated schematically in Fig. 48, the carbon dioxide flux has to overcome resistances additional to that experienced by water vapour, namely that encountered in the solution of carbon dioxide at the surface of the mesophyll cells lining the sub-stomatal cavity, and in its liquid-phase transfer to the chloroplasts located in the cytoplasm of cells. This liquid-phase resistance is often considerably greater than the gas-phase resistance r_{bl} and r_s (Slatyer and Denmead[28]). Carbon dioxide transfer from the atmosphere to plant leaves is further reduced by the concentration gradient often being lower than for water vapour.

Although it has been the purpose of this chapter to give a brief physical interpretation of some of the processes involved in plant–water relationships, it need hardly be added that different and wider concepts are essential even to a simple understanding.

Bibliography

1. KRAMER, P. J., Physical and physiological aspects of water absorption, in *Encylopaedia of Plant Physiology*, Vol. III (ed. W. Ruhland), p. 124. Springer-Verlag, Berlin, 1956.
2. MEES, G. C., and WEATHERLEY, P. E., The mechanism of water absorption by roots II. The role of hydrostatic pressure gradients across the cortex, *Proc. Roy. Soc.* **147B**, 381 (1957).
3. SLATYER, R. O., Internal water relations of higher plants, *A. Rev. Pl. Physiol.* **13**, 351 (1962).
4. VAN DEN HONERT, T. H., Water transport in plants as a catenary process, *Disc. Faraday Soc.* **3**, 146 (1948).
5. PHILIP, J. R., The physical principles of soil water movement during the irrigation cycle, *3rd Cong. Intern. Comm. on Irrigation and Drainage*, **8**, 125 (1957).
6. GARDNER, W. R., Dynamic aspects of water availability to plants, *Soil Sci.* **89**, 63 (1960).
7. DENMEAD, O. T., and SHAW, R. H., Availability of soil water to plants as affected by soil moisture content and meteorological conditions, *Agron. J.* **54**, 385 (1962).
8. LEMON, E., Energy and water balance of plant communities, in *Environmental Control of Plant Growth*. Academic Press, New York, 1963.
9. SLATYER, R. O., The significance of the permanent wilting percentage in studies of plant and soil water relations, *Bot. Rev.* **23**, 585 (1957).
10. GARDNER, W. R., and EHLIG, C. F., Some observations on the movement of water to plant roots, *Agron. J.* **54**, 453 (1962).
11. BROWN, H. T., and ESCOMBE, F., Static diffusion of gases and liquids in relation to the assimilation of carbon and translocation in plants, *Philos. Trans.* Series B, **193**, 223 (1900).
12. MASKELL, E. J., The relation between stomatal opening and assimilation, *Proc. Roy. Soc.* B **102**, 488 (1928).
13. PENMAN, H. L., and SCHOFIELD, R. K., Some physical aspects of assimilation and transpiration, *Symp. Soc. exp. Biol.* No. 5, 115 (1951).
14. HEATH, O. V. S., The water relations of stomatal cells and the mechanisms

of stomatal movement, in *Plant Physiology*, Vol. II (ed. F. C. Steward), p. 193. Academic Press, New York, 1959.

15. MILTHORPE, F., Plant factors involved in transpiration, in Plant-water relationships, *Arid Zone Research* **XVI**, 107. UNESCO (1962).

16. SLATYER, R. O., and BIERHUIZEN, J. F., Transpiration from cotton leaves under a range of environmental conditions in relation to internal and external diffusive resistance, *Aust. J. Biol. Sci.* **17**, 115 (1964).

17. SLATYER, R. O., BIERHUIZEN, J. F., and ROSE, C. W., A porometer for laboratory and field operation. *J. Exp. Bot.* **16**, 182 (1965.)

18. VAN BAVEL, C. H. M., NAKAYAMA, F. S., and EHRLER, W. L., *Measuring Transpiration Resistance of Leaves*. U.S. Water Cons. Lab. W.C.C. Report 2, Tempe, Arizona, 1964.

19. MONTEITH, J. L., Gas exchange in plant communities, in *Environmental Control of Plant Growth*. Academic Press, New York, 1963.

20. TAYLOR, S. A., and SLATYER, R. O., Water–soil–plant terminology, *Trans. 7th Internat. Congress of Soil Science, Madison*, **1**, 394 (1960).

21. SLATYER, R. O., and TAYLOR, S. A., Terminology in plant- and soil–water relations, *Nature, Lond.* **187**, 922 (1960).

22. BOLT, G. H., and FRISSELL, M. J., Thermodynamics of soil moisture, *Netherland J. Agric. Sci.* **8**, 57 (1960).

23. International Society of Soil Science, *Soil Physics Terminology*, Bulletin No. **22**, 5 (1963).

24. DAINTY, J., Ion transport and electrical potentials in plant cells, *A. Rev. Pl. Physiol.* **13**, 379 (1962).

25. CURTIS, O. F., and CLARK, D. G., *An Introduction to Plant Physiology*. McGraw-Hill, New York, 1950.

26. DAINTY, J., Water relations of plant cells, in *Advances in Botanical Research*. Academic Press, New York, 1963.

27. KEDEM, O., and KATCHALSKY, A., A physical interpretation of the phenomenological coefficients of membrane permeability, *J. Gen. Physiol.* **45**, 143 (1961).

28. SLATYER, R. O., and DENMEAD, O. T., Water movement through the soil–plant–atmosphere system, *Proceedings of the National Symposium on Water Resources, Use and Management*, Canberra, C2, 1963.

29. PHILIP, J. R., Osmosis and diffusion in tissue: half-times and internal gradients, *Plant Physiol.* **33**, 275 (1958).

Some Physical Constants and Conversion Factors

Physical Constants

Stefan's constant (or the Stefan–Boltzmann constant)

$$\sigma = 5\cdot67 \times 10^{-5} \text{ erg cm}^{-2} \text{ sec}^{-1} \text{ deg}^{-4} \text{K}.$$
$$= 8\cdot13 \times 10^{-11} \text{ cal cm}^{-2} \text{ min}^{-1} \text{ deg}^{-4} \text{K}.$$

Gas constant $R_u = 8\cdot31 \times 10^7 \text{ erg mole}^{-1} \text{ deg}^{-1}.$

Mechanical equivalent of heat $= 4\cdot18 \times 10^7 \text{ erg cal}^{-1}$
$$= 4\cdot18 \text{ joule cal}^{-1}.$$

Standard atmospheric pressure $= 1\cdot013 \times 10^6 \text{ dyne cm}^{-2}.$

Illustrative Values of Some Physical Properties

The values of the physical properties given below are all dependent on temperature and some depend on pressure. The values given apply only to the conditions noted in brackets.

Water

Latent heat of vaporization of water
$$= 586 \text{ cal g}^{-1} \quad (20°C)$$
$$= 2450 \text{ joule g}^{-1} \quad (20°C).$$

Dynamic viscosity of water
$$= 1\cdot00 \times 10^{-2} \text{ poise} \quad (20°C).$$

Thermal diffusivity
$$= 1\cdot43 \times 10^{-3} \text{ cm}^2 \text{ sec}^{-1} \quad (20°C)$$
$$(= \text{thermal conductivity numerically}).$$

Surface tension of water against air
$$= 73 \text{ dyne cm}^{-1} \quad (20°C).$$

Water vapour

Diffusivity of water vapour in air
$$= 0\cdot257 \text{ cm}^2 \text{ sec}^{-1} \quad (20°C)$$

Air

Specific heat at constant pressure

$$= 0.242 \text{ cal g}^{-1} \text{ deg}^{-1} \text{ C} \quad (20°\text{C}).$$

Thermal conductivity

$$= 6.1 \times 10^{-5} \text{ cal cm}^{-1} \text{ sec}^{-1} \text{ deg}^{-1} \text{ C} \quad (20°\text{C}).$$

Thermal diffusivity

$$= 0.21 \text{ cm}^2 \text{ sec}^{-1} \quad (20°\text{C}).$$

Density

$$= 1.2 \times 10^{-3} \text{ g cm}^{-3} \quad (20°\text{C}, 76 \text{ cm Hg}).$$

Dynamic viscosity

$$= 0.18 \times 10^{-3} \text{ poise} \quad (20°\text{C}).$$

Conversion Factors

(i) Length units:

$$\text{Inch (in.)} = 0.0254 \text{ m.}$$
$$\text{Micron } (\mu) = 10^{-6} \text{ m.}$$
$$\text{Ångström (Å)} = 10^{-10} \text{ m.}$$

(ii) Mass units:

$$\text{Pound (lb)} = 0.454 \text{ kg.}$$

(iii) Pressure or stress units:

$$1 \text{ lb ft}^{-2} = 4.88 \text{ kg m}^{-2}.$$
$$1 \text{ mb} = 10^3 \text{ dyne cm}^{-2}$$
$$= 0.75 \text{ mm Hg}$$
$$= 0.0295 \text{ in. Hg.}$$

(iv) Energy:

$$1 \text{ erg} = 10^7 \text{ joule.}$$
$$1 \text{ B.T.U.} = 1055 \text{ joule}$$
$$= 252 \text{ cal.}$$

(v) Power:

$$1 \text{ horse power} = 550 \text{ ft lb sec}^{-1}$$
$$= 746 \text{ watt.}$$
$$1 \text{ watt} = \text{joule sec}^{-1}.$$

(vi) Temperature:

$$t°\text{F} = 5(t - 32)/9°\text{C.}$$
$$T°\text{K} = t°\text{C} + 273.$$

(vii) Energy flux density:

$$1 \text{ cal cm}^{-2} \text{ min}^{-1} = 1 \text{ langley min}^{-1}$$
$$= 69.8 \text{ milliwatt cm}^{-2}.$$

1 mm hr^{-1} (evaporation of water, based on a latent heat of vaporization of 590 cal g^{-1})

$$= 59 \text{ cal cm}^{-2} \text{ hr}^{-1}$$
$$= 68.9 \text{ milliwatt cm}^{-2}.$$

Author Index

The numbers in italics refer to the page in the Bibliography in which the reference appears.

221

Subject Index

225